SYRIANUS

On Aristotle

Metaphysics 3-4

SYRIANUS

On Aristotle Metaphysics 3-4

Translated by
Dominic O'Meara and John Dillon

B L O O M S B U R Y
LONDON • NEW DELHI • NEW YORK • SYDNEY

Bloomsbury Academic
An imprint of Bloomsbury Publishing Plc

50 Bedford Square
London
WC1B 3DP
UK

1385 Broadway
New York
NY 10018
USA

www.bloomsbury.com

Bloomsbury is a registered trade mark of Bloomsbury Publishing Plc

First published in 2008 by Gerald Duckworth & Co. Ltd.
Paperback edition first published 2014

British Library Cataloguing-in-Publication Data
A catalogue record for this book is available from the British Library.

ISBN HB: 978-0-7156-3665-7
PB: 978-1-4725-5817-6
ePDF: 978-1-4725-0153-0

Library of Congress Cataloging-in-Publication Data
A catalog record for this book is available from the Library of Congress.

Acknowledgements
The present translations have been made possible by generous and imaginative funding from the following sources: the National Endowment for the Humanities, Division of Research Programs, an independent federal agency of the USA; the Leverhulme Trust; the British Academy; the Jowett Copyright Trustees; the Royal Society (UK); Centro Internationale A. Beltrame di Storia dello Spazio e del Tempo (Padua); Mario Mignucci; Liverpool University; the Leventis Foundation; the Arts and Humanities Research Council; Gresham College; the Esmée Fairbairn Charitable Trust; the Henry Brown Trust; Mr and Mrs N. Egon; the Netherlands Organisation for Scientific Research (NOW/GW); Dr Victoria Solomonides, the Cultural Attaché of the Greek Embassy in London. The editor wishes to thank Pieter d'Hoine, Christoph Helmig and Carlos Steel for their comments, Michael Griffin for preparing the volume for press, Fiona Leigh for providing comments to the author and proofreading with the assistance of Sarah Francis, and Deborah Blake at Duckworth, who has been the publisher responsible for every volume since the first.

Typeset by Ray Davies

Contents

Conventions

[...] Square brackets enclose words or phrases that have been added to the translation or the lemmata for purposes of clarity, as well as those portions of the lemmata which are not quoted by Syrianus.

<...> Angle brackets enclose conjectures relating to the Greek text, i.e. additions to the transmitted text deriving from parallel sources and editorial conjecture, and transposition of words and phrases. Accompanying notes provide further details.

(...) Round brackets, besides being used for ordinary parentheses, contain transliterated Greek words and Bekker page references to the Aristotelian text.

{...} Braces contain words which the editors regard as added later to the text that Syrianus wrote.

Introduction

1. Aristotle's *Metaphysics* in the Neoplatonic schools

When the young Proclus came to Athens in 430/431 to study philosophy, he began his instruction under Plutarch, then the head of the Platonist school there. However, the aged Plutarch soon died, to be succeeded in 432 by Syrianus.[1] Syrianus provided Proclus with a course of instruction that reflected the curriculum prescribed in the Platonist schools of Athens and Alexandria in the fifth and sixth centuries. In telling us about Proclus' course of study under Syrianus (*Life of Proclus,* ch. 13), Marinus informs us that in less than two years (in 432-434) Proclus read all of Aristotle's treatises with Syrianus, a reading which included logic, ethics, politics, physics and 'the science above these, theological science [i.e. metaphysics]' (*VP* 13,1-4). We can thus imagine the young Proclus reading Aristotle's *Metaphysics* with Syrianus along the lines suggested in Syrianus' own *Commentary on the Metaphysics*. Proclus was later to become, on Syrianus' death in 437, the head of the Athenian school and he would in his turn train, among others, Ammonius, who became head of the Platonist school in Alexandria and whose lectures on Aristotle's *Metaphysics* would be recorded by his pupil Asclepius in his *Commentary on the Metaphysics*.

Syrianus' two-year reading course in Aristotle covered, as Marinus suggests, a range of philosophical disciplines beginning with logic and progressing upwards from ethics and physics to the highest science, metaphysics. The whole Aristotelian course is described by Marinus as 'minor mysteries' preparing the initiation of the student into the 'major mysteries', i.e. a reading course in Plato's dialogues.[2] Proclus himself tells us a good deal about the 'major mysteries' and in particular that this course of reading had been introduced by Iamblichus.[3] It consisted of two cycles, a first cycle consisting of ten dialogues of Plato and a second cycle made up of two dialogues. The cycle of ten dialogues covered the same range, in the same progressive order, as the Aristotelian course, going from practical (i.e. ethical and political) sciences up to the theoretical sciences (i.e. physics and metaphysics). Metaphysics represented the highest science, the culminating point of the curriculum believed to be found in Plato's *Philebus*. In the second cycle, the two highest sciences were studied further, physics in Plato's *Timaeus*, and metaphysics in Plato's *Parmenides*.

If the second part of Proclus' training with Syrianus, the reading course in Plato, or 'major mysteries', had already been introduced by Iamblichus more than a century earlier in his school in Apamea in Syria, we may wonder if this is the case also for the first part of Proclus' training, the course in Aristotle, or 'minor mysteries'. If this were so, it would mean that already in Iamblichus' school Aristotle's *Metaphysics* was read as part of a highly structured philosophical curriculum. This would in turn indicate that Syrianus' *Commentary on the Metaphysics* stands in a tradition going back to Iamblichus. However, if Iamblichus' role in developing the course in Plato (the 'major mysteries') is well attested, we do not have, it seems, decisive evidence as to the origin of the course in Aristotle. Certainly, Aristotle's *Metaphysics* was read and discussed already in Plotinus' school in Rome in the 250s-270s (see Porphyry, *Life of Plotinus* 14,6-7). Plotinus' pupil Porphyry was also very active in incorporating Aristotelian texts, in particular Aristotle's logic, in the teaching of philosophy. And Iamblichus himself produced commentaries on (at least parts of) Aristotle's logic and, possibly, physics.[4] However, what indications have been found[5] of a commentary by Iamblichus on Aristotle's *Metaphysics* are not absolutely convincing. Yet it is not unlikely that the curriculum of 'minor mysteries' offered by Syrianus goes back, with the curriculum of 'major mysteries', to Iamblichus. The notion that Aristotle should be read *before* Plato, as a prior initiation, can be found in Themistius, well before Syrianus.[6] Iamblichus himself makes use of Aristotle's *Metaphysics*.[7] And in particular, the elaborate curricular theory that inspires Iamblichus' course of reading in Plato is anticipated in the course of reading in Aristotle. Thus the whole curriculum may well have had the same author, Iamblichus.

The information we have about the reading course followed by Proclus under Syrianus' direction is not only of interest as regards the presence of Aristotle's *Metaphysics* in the Neoplatonic schools in Late Antiquity. It also suggests why attention was given to the *Metaphysics* in these schools and what their approach to Aristotle's text would have been, if we also take into account what we otherwise know about the later Neoplatonic philosophical curriculum.[8] Syrianus took Proclus through Aristotle's works, as noted above, in relation to a range of philosophical disciplines and according to a progressive order, the same range and order as that inspiring the course of reading in Plato that Proclus would then follow. This range included the practical sciences (preceded by logic) which were followed by the theoretical sciences, physics and metaphysics. These sciences, in this order, represented an ascending scale of knowledge, going from practical knowledge engaged in material affairs to theoretical knowledge of the universe, a knowledge bringing the student to the discovery of the transcendent, immaterial, divine causes of the universe, a discovery which facilitated the transition to the philosophical science of the divine, 'theological' science, or metaphysics, the highest form of knowledge attained in the curriculum. Metaphysics was thus the goal of the curriculum, reached, at a preparatory level and (presumably) to an inferior degree, by reading Aristotle's *Metaphysics*, and, at a superior level, far more adequately, we may suppose, by a reading of Plato's *Parmenides*, the culmination of the course in Plato's dialogues and of the curriculum as a whole. If knowledge of the divine is consequently the goal of the curriculum, this fits well with the way in which later Neoplatonists defined the goal of philosophy, as 'assimilation to God to the extent possible' (Plato, *Theaetetus* 176B).[9] Assimilation to God means, for humans, living at the highest level of life possible to humans, the life of the divine in humans, the life of reason, of which knowledge of the divine represents the highest point.

These ideas, for which Neoplatonists could find support not only in Plato but also in Aristotle's *Metaphysics* (book 1, chs 1-2) and *Nicomachean Ethics* (book 10, ch. 7), suggest that the reason why Aristotle's *Metaphysics* was of interest to later Neoplatonists was that the treatise was thought to convey knowledge of divine being and to contribute thus to the assimilation of the soul to the divine. However, the position of the *Metaphysics* in the curriculum, as part of a *preliminary* course, preparing access to Plato, also indicates that Aristotle's treatise was not considered to be a fully satisfactory, complete text in the science of the divine: such a text was, in their view, Plato's *Parmenides*. Aristotle's treatise was thus both pre-eminent, as the culmination of the preliminary curriculum, and subordinate, precisely as part of this preliminary curriculum. The curriculur position of Aristotle's *Metaphysics* thus has implications for the way in which the treatise was understood by later Neoplatonists, the supposed purpose of the treatise, its contents, methods and limitations. We may see this in more detail by considering the explanation Syrianus himself gives in his *Commentary on the Metaphysics* of how he wishes to use Aristotle's *Metaphysics*.

2. The purpose and extent of Syrianus' commentary

The manuscript tradition has transmitted Syrianus' commentary on books 3, 4, 13 and 14 of Aristotle's *Metaphysics*. We may consequently wonder why we have his commentary only on these four books of the *Metaphysics*. A number of options can be envisaged: (1) might there have originally existed a complete commentary of which only some parts have survived due to the vicissitudes of manuscript tradition?; (2) or did Syrianus produce a commentary only on books 3, 4, 13 and 14 of the *Metaphysics* (and why just on these books)?; (3) or might his commentary have been somewhat more extensive than which presently survives, yet not covering all of Aristotle's treatise?

The prefaces which Syrianus gives to his commentary on books 13-14, on book 3 and on book 4 of the *Metaphysics* point rather in the direction of options (2) or (3). In the preface to books 13-14, while praising Aristotle as a philosophical benefactor of humanity, in particular as regards logic, ethics and physics (i.e. sciences relating to the material world), Syrianus finds Aristotle seriously lacking in regard to theological science (metaphysics), to the extent in particular that he attacks the metaphysics of the Pythagorean-Platonic tradition, especially in *Metaphysics* books 13 and 14. Aristotle's criticisms of Pythagorean-Platonic metaphysics, Syrianus suggests, could have the effect of inducing in the student a contempt for 'divine realities and the inspired philosophy of the ancients' (80,23-5). It is therefore necessary to show that Aristotle's criticisms fail and this is what Syrianus undertakes to do in his commentary on *Metaphysics* books 13-14. He also remarks at the end of this commentary (195,10-12) that Aristotle expresses the same criticisms elsewhere in the *Metaphysics*, notably in book 1, and thus that the refutation of the criticisms in books 13-14 also applies to the criticisms to be found in book 1. In his preface to book 3, Syrianus describes how Aristotle here develops arguments in support of opposing answers to a series of questions concerning the scope of metaphysics: Syrianus will show in his commentary what the right (i.e. Platonic!) answers are and why the arguments Aristotle gives in support of the opposite answers are not valid (1,20-2,3).[10] Finally, in the preface to book 4, Syrianus indicates that Alexander of Aphrodisias provides adequate explanation of Aristotle's text in his *Commentary on the Metaphysics*: Syrianus will limit himself to providing a paraphrase of the text, so as to preserve its coherence, while discussing some difficulties in it (54,12-15).

From these indications we can infer that it is Syrianus' purpose to compensate for the metaphysical deficiencies of Aristotle's *Metaphysics*: to refute Aristotle's criticisms of Pythagorean-Platonic metaphysics and to show where Aristotle's arguments against Pythagorean-Platonic answers to metaphysical questions fail. This is necessary insofar as a student might be led astray by Aristotle's text and have contempt for (Pythagorean-Platonic) metaphysics. If this is the purpose of Syrianus' commentary on books 13-14 and 3 of the *Metaphysics*, in his commentary on book 4 Syrianus appears to have a somewhat different approach. In book 4 Aristotle provides answers to major questions concerning the scope of metaphysics as a science: Syrianus largely agrees with Aristotle's views and provides, in his paraphrase, an overview of what metaphysics is, while discussing some difficulties, which, we might suppose, could include (but not be restricted to) issues separating Platonism and Aristotelianism such as the question of the relation between unity and being (see 59,3ff.). Otherwise, for the actual explanation of Aristotle's text, Alexander of Aphrodisias' commentary suffices. In short, as has been noted,[11] Syrianus' *Commentary on the Metaphysics* is not a commentary on the *Metaphysics* in the sense of a continuous explanation of the text such as that provided by Alexander of Aphrodisias. Syrianus' work is rather a corrective, or 'antidote' (or perhaps a kit of antidotes!), to be used by the student who reads Aristotle's work; for actual explanations of passages in the text, the student can use Alexander of Aphrodisias' commentary. The student will also find in Syrianus' commentary on book 4, not a commentary (such as Alexander's), but an overview of metaphysics.[12] The metaphysical deficiencies that Syrianus notes in Aristotle's treatise are expressed in the distinction he sometimes makes[13] between the 'demonic' Aristotle and the 'divine' Platonists and Pythagoreans: this indicates a subordinate rank in philosophical insight, as demons are subordinate to gods.

From these observations we can conclude that no new complete commentary on Aristotle's *Metaphysics* is required in the philosophical curriculum: Alexander of Aphrodisias' *Commentary* will suffice for explaining the text. However, Aristotle's treatise and Alexander's commentary should be accompanied by refutation, where they attack Pythagorean-Platonic metaphysics or take positions against it, a refutation supplied by Syrianus. It therefore seems unlikely that Syrianus saw any need to produce a commentary on every book of Aristotle's *Metaphysics*. However, Aristotle criticises Pythagoreans and Platonists elsewhere in his treatise, not just in books 13-14 (and 1), and it may be that Syrianus' 'antidote' extended somewhat further than what we now have: there are indications in Ammonius' course on the *Metaphysics*, as recorded in Asclepius' *Commentary*, of Syrianus' discussion of particular passages in *Metaphysics* book 7, where Aristotle again argues against Platonism.[14]

3. Syrianus' conception of metaphysics

In the preceding pages, mention has been made, for the purpose of simplicity of reference, of 'metaphysics' as the science with which Aristotle is concerned, in Syrianus' view, in the *Metaphysics*. Although 'metaphysics' as an expression is already used in late Antiquity,[15] it is not used by Syrianus in his commentary, who, following Aristotle's own terminology, refers rather to 'wisdom', 'first philosophy' or 'theology'. Thus Syrianus refers to Aristotle's *Metaphysics* as the 'theological treatise' (80,17; cf. 1,26). 'Theological [science]' is an expression used by Aristotle in an important passage in book 6 (ch. 1, 1026a19ff.) so as to indicate a philosophical science dealing with divine unmoved substance, a science which would be 'first philosophy', if there be such a substance, and which would concern in some way 'being as being', the object of the science Aristotle discusses in book 4. Probably inspired at least in part by this passage of book 6 of the *Metaphysics*, Syrianus takes it that the 'wisdom' Aristotle sketches at the beginning of book 1, the science of 'being as being' of book 4 and the 'first philosophy' or 'theological science' of book 6 are all descriptions of one and the same science, which Syrianus also associates (55,29-33) with the highest philosophical knowledge indicated by Plato in the *Republic*, the knowledge of Forms and of the Form of the Good named 'dialectic' (*Rep.* 510B-511D). Syrianus furthermore distinguishes between this Platonic 'dialectic', which is what he thinks Aristotle's wisdom-first philosophy-theological science aims at being, from a 'dialectic' which Aristotle himself (1004b17-26) makes subordinate to metaphysics and intermediate somehow between metaphysics and sophistry (see Syrianus' commentary, 2,25; 104,28-31).

Syrianus takes it then that there is one supreme philosophical science, (Plato's) dialectic or (Aristotle's) wisdom-first philosophy-theology, which the student can discover (imperfectly) in Aristotle's *Metaphysics* and (perfectly) in Plato's *Parmenides*. However, Aristotle's descriptions of the objects of a science (or sciences) in books 1, 3, 4 and 6 of his treatise pose notorious problems which could cast doubt on the possibility of there being *one* science of these objects. In what follows, it will be suggested briefly how Syrianus dealt with some of these problems and understood the systematic unity of the objects of metaphysical science and thus of the science itself. We will take the three problems selected by Syrianus in the overview he gives at the beginning of his commentary on book 4 (54,4-8).

One of the problems posed by Aristotle himself in book 3, the third problem, concerns whether there is one science of all substance, or different sciences for the different kinds of substance (997a15-16). Syrianus, in his comments on the passage (21,3-7), speaks of one supreme science which concerns all being as being, and of other sciences that deal with different parts (or modes) of being. It might thus appear that metaphysics is a science of all being, whereas 'theological' science would then be, we might infer, the science of a part of being, divine substance. However, Syrianus makes his position clear in his commentary on book 4 (57,22ff.). If metaphysics concerns all being, it deals in the first instance with that which primarily is, i.e. divine substance, that on which other beings depend and which gives them being. Syrianus describes the philosophical sciences, or parts of philosophy, not as structured in terms of a genus (the science of all being) and its co-ordinate species (specialised sciences dealing with different parts or modes of being), but as constituting a series ordered in terms of priority and posteriority, such that the first in the series, first philosophy, deals with prior substance (divine or intelligible substance), whereas the sciences that come after it in the series (mathematics and/or physics) deal with posterior substances, i.e. substances that depend for their being on divine intelligible substance (58,12-19; 61,17-28).[16] Thus theological science will be the science of all being, since it deals with what primarily is, divine intelligible being, which gives being to whatever else is. This being is, for Syrianus, the transcendent Platonic Forms (the objects of dialectical knowledge in Plato's *Republic*), which are, in his view, the content of the thought of what Aristotle in the latter part of *Metaphysics* book 12 identifies as divine substance, a transcendent Intellect.

A further problem raised by Aristotle in book 3, the fifth problem, concerns what he calls the essential attributes or accidents of being, such as unity/multiplicity, sameness/difference, likeness/unlikeness: do they also come within the scope of the science of being? Syrianus thinks that they do, since they belong to all being and not just to one part of being. However, here again, these essential attributes are ordered in a series of prior and posterior terms, corresponding to the hierarchy of the levels of reality, such that the essential attributes are found in primary and in derivative forms (5,16-27). With these attributes Syrianus combines (5,27-33) the 'major kinds' of Plato's *Sophist* and the forms discussed in the second part of Plato's *Parmenides.*

Finally, one might note the second problem raised by Aristotle in book 3: does the science of being, which, as a science, will deal with the causes of being, also deal with the principles of demonstration, i.e. premises or axioms such as are used by mathematicians, which are presupposed by demonstrative arguments but not themselves proved by such arguments, the most important of such axioms being, Aristotle indicates, the Principle of Non-Contradiction. Syrianus, agreeing with Aristotle, thinks that such axioms, in relating to all being, cannot be studied in a specialised science, but must come within the range of the science of all being. However, Syrianus provides a further explanation of the necessary unity of the science of the causes of being and the science of the principles of demonstration: 'wisdom' (i.e. metaphysics), he suggests (2,29-30; 3,21-2; 19,5-6; 20,7-8), is the proximate product of a transcendent divine Intellect; as such, wisdom or first philosophy 'imitates Intellect, in which intellection (*noêsis*) and the intelligible (*noêton*) are not divided, [wisdom] being assimilated to intellection through the knowledge of axioms, and to the intelligible through knowing being' (20,29-31). Metaphysics is (or should be!) an image of a divine transcendent Wisdom, in which the act of thinking (intellection) is one with its objects (the intelligible). So human wisdom will unite knowledge of the principles of thinking (the axioms at the foundation of every demonstrative argument) with that of the causes of being.

The conception of metaphysics as a human science in the image of divine science can be elaborated further if we take account of Syrianus' theory of scientific knowledge.[17] A science such as mathematics may be aroused by perception of material reality, but it does not derive its universal and necessary truths from this. Universals abstracted from sense-perception (universals named by Syrianus as 'later-born', *hustero-genê*) cannot have the precision and necessity of scientific truths. These are found rather as 'reason-principles' (*logoi*) immanent in the human soul and brought to expression by thought (4,32-5,2; 24,4ff.; 90,4-16; 161,30-4). These reason-principles (also called 'intermediate substances', 4,33) are present in soul in virtue of the way in which soul, according to Plato's *Timaeus*, is structured by the divine maker of the world, or 'demiurge' (4,8-11). The demiurge uses the same reason-principles in structuring the world. Thus if the mathematician discovers in his thinking universal and necessary truths which turn out to govern the physical world, this is because the knowledge that the mathematician discovers in his soul is that which inspires the making of the world (see 27,30-7; 88,24-7). There are also more fundamental truths in the human soul, those concerning intelligible or divine substances and their first causes (see 27,36-7; 90,4-7): it is these truths that are articulated by the metaphysician, for example by Syrianus' pupil Proclus in his *Elements of Theology*.

The metaphysical reason-principles explicated by the metaphysician are open to the methods of scientific reasoning, definition, demonstration, analysis and synthesis, even if the objects to which they refer, transcendent Forms, divine Intellect, and their causes, are strictly speaking beyond the range of human scientific thinking and are known, not by scientific reasoning, but by non-discursive insight (4,24-37; 80,12-13; 100,28-9; 115,21-6; 147,14-15). It is in this way that there can be a scientific reasoned discourse about realities which are beyond such discourse in that their nature escapes scientific method.[18]

However, in his commentary, Syrianus is concerned, not so much with elaborating such a discourse, as does Proclus in his *Elements of Theology*, as with responding to Aristotle's deviations from metaphysical knowledge, for which purpose he refers, as if to a canon, to the doctrines and texts transmitted by the major Pythagoreans and Platonists: Pythagoras, Parmenides, Empedocles,[19] Plato himself, of course, Plato's interpreters, Plotinus, Porphyry and Iamblichus, all of whom, and contrary to Aristotle's insinuations, unanimously express in Syrianus' view, the same metaphysical doctrine, to be found also in the Orphics and the poetical 'theologians' cited by Aristotle. It is against this Pythagorean-Platonic tradition, taken as a canon, that Syrianus measures the strengths and deficiencies of Aristotle's *Metaphysics*.

The student, beginning the study of theological science, follows to some extent in the footsteps of the youthful Socrates who, in the second part of Plato's *Parmenides*, is subjected to dialectical training ('gymnastics', *Parm*. 135D-136A). Since the second part of the *Parmenides* was taken by the Neoplatonists as representing metaphysics, this meant that metaphysics could be approached first through preparatory metaphysical exercises or training. This idea is expressed in Proclus' *Commentary on the Parmenides*,[20] and it seems to be present also in Syrianus' approach (13,5; 30,34; 36,3; 40,10; 48,34) to the study of the opposing arguments concerning the scope of metaphysics set up in book 3 of the *Metaphysics*: here also the student could be prepared ('trained') for metaphysics, as a preliminary to the study of Plato.

4. Syrianus' sources in his commentary on *Metaphysics* 3 and 4.

From the preceding it can be seen that Syrianus reads Aristotle's *Metaphysics* with the help of different sources which he uses in two different ways: (1) for the explanation of passages in Aristotle, Syrianus makes use of Alexander of Aphrodisias' *Commentary on the Metaphysics*;[21] (2) however, where Alexander, as a convinced Aristotelian, sides with Aristotle's criticisms of Pythagorean-Platonic metaphysics or adds to these criticisms, Syrianus argues against Alexander (cf. 23,25ff.; 32,15). The refutation of Aristotle's and Alexander's criticisms of Pythagorean-Platonic metaphysical doctrines is based in Syrianus on the supposition that there is a coherent metaphysical theory to be found in the texts of the major Pythagoreans and Platonists. These texts will thus be cited against Aristotelian claims, texts taken from Pythagorean authors, from Plato, from Plato's (Neoplatonic) commentators, from Orphic and from other ancient theological poetry. However, the coherent metaphysical theory thought by Syrianus to be present in these texts is clearly Neoplatonic in inspiration and largely the creation of Iamblichus.

If we could be certain that Iamblichus had produced a commentary on the *Metaphysics*, we could suppose that Syrianus derived his metaphysical theory (and many of the ancient texts cited in support of it) from there. But Syrianus nowhere cites such a commentary. However, Syrianus could probably also have found his Pythagorean-Platonic metaphysical theory and texts in Iamblichus' *Commentary on the Parmenides*, a text which he does mention (38,38-9), as well as in other Iamblichean works, in particular the 10-volume treatise *On Pythagoreanism*.

The importance of Iamblichus' work *On Pythagoreanism*[22] for Syrianus becomes evident in his commentary on *Metaphysics* 13-14. At 103,6-7 he refers to the work as a whole, after having reproduced (in 101,29-102,35) many chapter-headings of book 3 of the work (*De communi mathematica scientia* 3,7-8,6) by way of an overview of Pythagorean philosophy of mathematics. We can also detect Syrianus' use of book 4 of Iamblichus' *On Pythagoreanism* (*In Nicomachi arithmeticam introductionem*) in a number of places.[23] Furthermore Syrianus refers to book 5 of Iamblichus work (*On Arithmetic in Physical Matters*) at 149,28-31 and can be seen, from fragments surviving from this book, to be using it elsewhere in his commentary.[24] Finally Syrianus refers to book 7 of *On Pythagoreanism* (*On Arithmetic in Theological Matters*) at 140,15. From these passages we can conclude that Syrianus derived from Iamblichus' work, not only the general lines of the metaphysical theory which he attributes to the Pythagorean-Platonic tradition, but also a number of the supposedly Pythagorean texts which he cites in support of this theory. Basically the same theory is presupposed in Syrianus' commentary on *Metaphysics* 3 and 4 (cf. for example 10,1-11). It is likely that Syrianus also made use of sources of information other than those noted here.

5. Notes on the translation

In the Introduction (part 5) of our translation of Syrianus' commentary on *Metaphysics* 13 and 14, indications are given concerning manuscripts and editions of Syrianus' commentary.[25] While awaiting the critical edition of the Greek text edited by Concetta Luna, annotated by Cristina d'Ancona and translated into French by Jean-Pierre Schneider, to be published by

Les Belles Lettres (Paris), the present English translation of Syrianus' *Commentary on Metaphysics 3-4, 13-14* follows the edition of the Greek text published in 1902 by W. Kroll, who reports in his critical apparatus readings of the principal manuscript, Paris Coislin 161 (= C),[26] as well as corrections suggested by Bagolinus' Renaissance Latin translation and by H. Usener and A. Brandis (in the Usener edition; see below Select Bibliography). For the present English translation of the commentary on *Metaphysics* 3-4, a microfilm of manuscript C has also been consulted and in particular very much help has been received from Schneider's (as yet unpublished) excellent annotated French translation which will accompany Luna's critical edition. In some pages the Greek text can be improved thanks to excerpts made from it by the Byzantine philosopher Michael Psellos in the eleventh century, well before our earliest surviving Greek manuscripts of Syrianus. The fact that Psellos used Syrianus' *Commentary on the Metaphysics* is also of interest in that Psellos' pupil and successor Michael Italos taught Michael of Ephesus. If Michael of Ephesus is indeed the author, as has been argued, of Pseudo-Alexander's *Commentary on the Metaphysics 6-14*,[27] then his use of Syrianus' commentary is anticipated by Psellos.[28]

In publishing his edition of Syrianus' commentary, Kroll (i) *abbreviated* the Aristotelian lemmata, which are almost always given[29] in full in manuscript C. The present English translation therefore includes the full text of the lemmata.[30] The Aristotelian lemmata appear to represent the text of Aristotle that Syrianus himself used and thus are of considerable interest for establishing Aristotle's text. The modern reader interested in Aristotle's text can find collations of textual variants found in the lemmata of Syrianus' commentary on books 13 and 14 in Luna, 2005. As for the lemmata included in his commentary on books 3 and 4, variants are noted in the translation given in the present volume,[31] as compared with the Greek text printed in W. Jaeger's edition of the *Metaphysics* (Oxford, 1957), which is presumably at present the most available.

In his edition of Syrianus' commentary Kroll furthermore (ii) *rearranged* some of the Aristotelian lemmata (following a typographical arrangement adopted already in the Usener/Brandis edition) so that they appear on the printed page *within* the commentary on a previous lemma. However, this typographical layout does not correspond to the presentation of the lemmata in manuscript C, where each lemma, with its commentary, simply follows on the preceding lemma and its commentary. Kroll's layout is moreover misleading in that it suggests that Syrianus' work is a continuous explanatory commentary on the text of Aristotle in the style of Alexander of Aphrodisias' commentary.[32] However, as noted above, Syrianus' text is not such a commentary: for the explanation of passages in Aristotle, Syrianus' reader may use Alexander' commentary; Syrianus will provide correct answers and refutation where Aristotle conflicts with Pythagoreanism-Platonism. The following English translation of Syrianus' discussion of *Metaphysics* 3 will not therefore follow

Kroll's typographical example and will simply print the lemmata with their respective comments in a continuous series.[33]

Finally, as regards the English translation offered here, the attempt has been made to stay as close as possible to the Greek. The usual dilemmas occur as regards translating the terminology of such a text in Greek metaphysics. In order to facilitate comparison with the English translation published by W. Dooley and A. Madigan (in the present series) of Alexander of Aphrodisias' commentary, to the extent possible the same (or closely related) translations of Greek terms as theirs have been adopted. As regards the notorious difficulties involved in translating certain fundamental terms, the following approach has been followed. *Ousia* has in general been translated as 'substance', to the extent that 'substance' may be less restrictive than 'essence' and can cover a range going from material substances (concrete individuals) to immaterial substances (whose substance is identical to their essence). 'Being' is used where *ousia* bears a broader meaning and does not refer to particular material or immaterial substances. With *eidos* it is sometimes impossible to use one single English equivalent. The Aristotelian text and Syrianus' corresponding comments oblige us sometimes to translate this word with 'kind' or 'species', whereas elsewhere the same Greek term can refer to 'forms' which may range from immanent (enmattered) forms to transcendent (Platonic) Forms. The continuity in the Greek terminology is important, since it reflects for Syrianus a metaphysical continuity (in the case where the 'kinds' or 'species' are natural, and not 'later-born' conceptual abstractions produced by human thought from sense-perception). Similar problems arise in translating *hen* or *to hen*: we have usually translated these as 'one' or 'unity'. However in some cases what is clearly meant by Syrianus is 'the One' as the metaphysical first principle, the principle of the unity of everything. Here, as in the case of *eidos*, the attempt has been made to capitalise with sobriety. In translating the Aristotelian lemmata, an effort has been made to avoid implicit interpretation of Aristotle's text, since what is of concern here is presumably not what Aristotle means in a passage, but what Syrianus makes of the passage.

We have inserted in our translation references to Aristotle's *Metaphysics*, where Syrianus is commenting on a specific passage, or referring or alluding to one. These references are given without mention of the name of the author or of the work, and take the form of a page number, column (a,b) and line number (e.g. 996a18), according to the standard practice. We follow the same practice in the notes to the translation, where references to other parts of Syrianus' commentary are also given, here too without the name of the author or work, consisting simply of (a) page and line number(s) (e.g., 64,3), according to Kroll's edition. In referring to the commentaries of Alexander of Aphrodisias and of Asclepius on the *Metaphysics*, we simply refer to 'Alexander', or 'Asclepius', followed by (the) page and line number(s) in Hayduck's edition of these works.

In this volume the Introduction, translation and notes have been done by DO'M. On some points (see nn. 27, 33) JD takes a different view. The translation has been checked and improved by JD and has much benefited from the unpublished French annotated translation by J.-P. Schneider, to whose *megalopsukhia* DO'M is much indebted. The translation has also been checked and improved by Carlos Steel and Christoph Helmig, who, with JD, have saved DO'M from a number of errors, if not perhaps (alas!) from all.

DO'M is also grateful for help received in preparing the volume from Michael Griffin, Luca Pitteloud and Euree Song.

Notes

1. On Syrianus' life and works, see our Introduction to *Syrianus: On Aristotle Metaphysics 13-14*, section 1 and Cardullo, 1995 (for the bibliographical details of the ancient and modern sources cited here and in what follows see the Select Bibliography below).

2. The image of philosophy as an initiation to mysteries (minor and major) comes from Plato, *Gorgias* 497C.

3. Proclus as reported in the [Anonymous] *Prolegomena to the Philosophy of Plato* 26,16-44; see the Introduction to the edition, lviii-lix and lxvii-lxxiii.

4. Larsen, 1972, 51-5.

5. See John Dillon's edition of the fragments of Iamblichus' commentaries on Plato, 22, and Steel, 1978, 124.

6. Themistius, *Or.* 20, vol. 2, 6,13-19 (cited in the edition of Marinus, *Life of Proclus*, 109 n. 2).

7. See for example Iamblichus, *Protrepticus* 20,2-6 (the scope of Aristotle's metaphysics in a nutshell) with the ancient scholion on 22,1ff. (127,16-19); also O'Meara, 1989, 77-8.

8. For more detail and references for what follows in this paragraph, see O'Meara, 2003, chs 4-6.

9. See Syrianus' commentary, below, 14,19-20.

10. Aristotle mentions the need to arbitrate between the opposing positions at the beginning of book 3 (995b3-4).

11. By Luna, 2004, 39.

12. The heterogeneous purposes of Syrianus' work may help explain its disparate character, both refutatory, corrective and paraphrastic.

13. See n. 32 below, p. 120.

14. For a study of the evidence for this see Cardullo, 1993a.

15. For example in Porphyry's *Life of Plotinus*, 14,6.

16. This kind of series is described as a 'quasi-genus' by A. Lloyd, 1990, ch. 3.

17. For more details on what is presented in this paragraph see O'Meara, 1986 and the Introduction to our translation *Syrianus: On Aristotle Metaphysics 13-14*, section 2.

18. Thus the transcendent One is beyond the Principle of non-contradiction (cf. 18,25-7).

19. For Parmenides and Empedocles as Pythagoreans, see 11,35-6; 43,8; 60,6-7.

20. See O'Meara, 2000, 282 and Syrianus 39,35-40,1; 47,35 (Zeno).

21. See Luna, 2001, Étude II, for a listing of passages.

22. Also referred to variously as the 'Pythagorean Sequence', or as the 'Summary of Pythagorean Doctrines', in the notes to *Syrianus: On Aristotle Metaphysics*

13-14. On the title and contents of Iamblichus' *On Pythagoreanism*, as well as on its influence on Syrianus see O'Meara, 1989, chs 2, 3, and 6.

23. Compare, for example, 140,6-10 with *in Nicom.* 11,1-9; 142,15-25 with *in Nicom.* 10,12-24; 165,13-14 with *in Nicom.* 6,20-2.

24. The fragments from book 5 are printed with a translation in O'Meara, 1989, 218-23. Compare Syrianus 149,28-31 with Iamblichus *On Arith. in Phys.* 222,90-1; Syrianus 130,34 with Iamblichus at 220,48-9; Syrianus 143,6-9 with Iamblichus at 220,49-53.

25. A detailed account of the manuscripts, in particular of MS C, can now be found in Luna, 2007.

26. Mid-fourteenth century; on its scribe, identified as Neophytos Prodromenos, cf. Luna, 2007.

27. The case for Michael of Ephesus has been argued by Luna, 2001, Étude I, in my view convincingly. For another view see the Introduction to *Syrianus: On Aristotle's Metaphysics 13-14*, section 3.

28. Another later Byzantine use of Syrianus (*c.* 1336, roughly contemporary wih MS C) is detected by R. Sinkewicz, 'The Solutions addressed to George Lapithes by Barlaam of Calabria and their philosophical context', *Medieval Studies* 43 (1981), 176-81.

29. But see 40,24, with n. 243 (below).

30. In some cases Syrianus covers passages in Aristotle going beyond the range of text given in the corresponding lemma (see for example 14,15ff.). It is possible then that some lemmata were omitted in MS C or in its MS sources (see Luna, 2007, 129, for an example at Syrianus 116,1). Or perhaps in some cases Syrianus could feel free to go further in his comments. A special problem is presented by Syrianus' commentary on book 4: only some brief initial lemmata are preserved in MS C. Does this mean that many lemmata have been lost (cf. Luna, 2007, 126-7)? One should also recall that Syrianus here wishes to paraphrase Aristotle, which would mean less need for lemmata.

31. But I have not noted where MS C sometimes omits a word or inverts word-order.

32. On this see the detailed analysis in Luna, 2004, 40-52.

33. Kroll's arrangement is followed, however, by John Dillon in the translation of the commentary on books 13 and 14.

Textual Emendations

The following list gives the places where our translation diverges from the Greek text printed in Kroll's edition, places where we adopt emendations (suggested in many cases by Kroll himself, or by Usener) noted in Kroll's critical apparatus but not included in the Greek text, or where we suggest our own emendations, or follow those mentioned by J.-P. Schneider in his unpublished French translation to accompany the forthcoming critical edition of the text prepared by C. Luna, in collaboration with C. d'Ancona.

2,20: Accepting Usener's suggestion *prôtôn* for *ontôn*.
6,30: Accepting Kroll's suggestion to read *stasin katekhein to* for *stasin kai* (but Kroll would add also *holon* after *to*).
7,3: Supplying *stoikheia* after *autou* from Psellos' excerpts.
7,10: Supplying *oute* before *hosa* from Psellos' excerpts.
7,17: After *toutôn* supplying *kai huper tauta tas dêmiourgikas ideas arkhas, hai kai*, from Psellos' excerpts (who adds *tithentai* after *arkhas*), replacing *eti*; this remedies the lacuna marked by Kroll.
9,25: *epi de* we correct to *epi <ta> de*.
16,28: Moving *hêmas* backward a few words from its present position, so as to be taken with *hikanon eipein*.
20,9: Adopting Usener's suggestion to read *kata panta* for MS C's *katatakta*.
20,17: Adopting Usener's suggestion *tou ... epeskemmenou* for *to ... epeskemmenon*.
22,22: Kroll mistakenly prints *khalepon* for *pankhalepon*.
23,18: Adopting Kroll's suggestion to read *kai <di'> auto*.
24,15: Adopting Usener's suggestion to read *<di>exodikas*.
27,33: Accepting Kroll's suggestion to correct *ta* to *tôn*.
28,19: Keeping *pros* in MS C, which Kroll corrects to *pro*.
30,24: Adopting Usener's correction of *genêtikai* to *genikai*.
33,22: Adopting Usener's suggestion to read *touto kata tou* for *kai tou*.
34,10: Following Kroll's suggestion to fill the lacuna with *ton de sullogismon*.
35,29: We correct the second *kai* to *ei*.
35,32: Reading *tauta* with Usener for *taûta*.
36,21: Following Kroll's suggestion to read *de tois epistêmosin tês ... aitias* for *de autois ekhontôn tên tês ... aitian*.

39,35: Following Kroll's suggestion to add *ou* after *heautou*.

41,18-19: Returning to MS C's version of the quote from Plato, *hin' oun thnêton te kai athanaton deontôs hapan*, which Kroll corrects to match Plato's text.

48,5: Suppressing *pros* following Bagolinus.

50,17: Adopting Usener's suggestion *ho logos* <*toioutos*>.

50,19: Reading with Usener *toiouto* for *touto*.

50,20: Keeping *ouk*.

50,23: Reading *diaireseis*, along the lines suggested by Usener; Kroll prints *energeiai* and suggests reading *tina*.

51,24: Following Usener's suggestion to supply <*hoion tessara tôn pantôn rhizômata*> before *phêsin*.

53,2: Following Usener's suggestion to supply here such words as <*holôs ouden leipetai gnôston, eite gignôskontai*>.

54,5: Supplying <*estin epistêmês eidenai*> with Usener; Kroll adds *tês* after *eidenai*.

54,14: Supplying <*axion*> before *exetaseôs* with Usener.

56,36: Reading *gennêtika* with Aristotle; MS C has *poiêtika* and Kroll prints *poiêta*.

58,19: Correcting *deuterô* to *B*, following Luna.

59,9: we read *huperousiou*; MS C has *huperousia*; Kroll prints *huper ousian*.

59,17: Deleting *mê*, as suggested by Kroll.

59,18: Correcting *homôs* to *holôs* as suggested by Kroll.

64,8: reading *sunônumôs* with MS C, as suggested by Schneider, which is changed to *homonumôs* by Usener and Kroll.

67,9: Replacing *nemontos* with *onoma* as suggested by Kroll.

69,7: Supplying <*kai ei pseusetai hê apophasis, alêtheusei hê kataphasis*> as suggested by Usener; Kroll marks a lacuna.

70,24: Keeping *horôsi* (MS C); Kroll prints *eôsi*.

72,2: Reading *didoasi ti kai hôristhai* with Usener; Kroll prints *didoasi te kai hôristai*.

72,8: Following Usener's suggestion to read *theseôn*; Kroll prints *diaire-seôn*.

74,33: Reading *diathrêsas* as suggested by Usener; Kroll prints *diakathêras*.

75,32: Reading *homôs* as suggested by Kroll, who prints *homoiôs*.

77,29 Keeping *prosêkanto*; Kroll corrects this to *prosêlanto*.

79,20: Suppressing the <*ou*> added by Kroll.

SYRIANUS

On Aristotle

Metaphysics 3-4

Translation

Syrianus, son of Philoxenus

On the Problems raised in a verbal[1] way in Book 3 of Aristotle's *Metaphysics* and deemed in need of arbitration

Preface[2]

In book 1 Aristotle says what the present investigation is, that it concerns being as being, which science it is, that it is first philosophy, how many causes there are, and how his predecessors discussed the causes of whatever in some way is. Then in book 2 he shows that these causes are not infinite 'either as a series or in kind' (994a2). Now, beginning book 3, he says (995a25) that we should first raise problems (*aporêsai*), so as to deal with greater chance of success with the topics at hand (for it is not possible to find a way forwards if the problems are not resolved, nor can they be resolved if we do not grasp what is problematic in the matter), and also so that we might see the goal of the present study. For the solution of the problems, in opening the way, will enable us to reach the most veritable goal of study (*theôria*).[3] He also shows as follows that it is right for a philosopher to raise problems: by becoming as it were an unprejudiced and impartial arbitrator of opposing arguments, the philosopher will choose what fits better with true knowledge (*theôria*).

Following these prefatory remarks, Aristotle first presents the problems (*aporiai*), after which he elaborates the difficulty of these problems, for each of them providing support for each side of the opposing arguments.[4] But one should not look for his solutions in this book, since the whole of book 3 is concerned with exploring difficulties. We, however, will both endeavour briefly[5] to give answers to the problems he presents, and, as regards the opposing arguments that he puts as questions, we will agree with some as being true, and combat others as being sophistical. For if such a goal is reached, then both the words (*lexis*) of the philosopher will be explained and the intention (*prohairesis*) of the treatise as a whole will be grasped in brief. For Aristotle will discuss, in his teachings [in this treatise] which go beyond nature,[6] problems which he has presented here and not others, save incidentally. Furthermore, the refutation of eristic 2,1 arguments will become easier and we will see how a Pythagorean and a Platonist might counter these objections. Now for the first problem.

Chapter 1[7]
First Part[8]
Presentation of the Problems

First Problem

995b5-6 Whether the causes are studied by one or by many sciences.

The question he is raising can be described as follows: since a cause may be efficient, formal and final, and, in some cases, material, is it the case that knowing something means knowing all its causes, there being one knower, as having one science; or will there be many sciences dealing with the many [kinds of] causes, one knowing the efficient cause, another, for example, the material or the final cause, another the formal cause?

We answer that one science examines all causes. For if it is ignorant of one of these, then it will not have knowledge of the thing. And if indeed the discussion be about wisdom (*sophia*), then it would be absolutely ridiculous to think that wisdom would be ignorant of the Good, or of form, or of the cause which generates beings.

Second Problem

995b6-8 And whether the science considers only the first causes of substance, or also the principles from which all [men] make demonstrations.

The question here is if he who does first philosophy[9] and knows the causes of being will also know the truth in premises (*protaseis*) that are most primary and grasped immediately (*amesois*); or if it is one person who knows the principles of being, and another who knows axioms (*axiômata*) that are primary[10] and most general.

But it is clear that the same person will know both the former and the latter. For if he is ignorant of either, he will not be perfect (*teleios*)[11] and he will not have, in relation to knowing each of both kinds of principles, the science of sciences.[12] [This science involves] him knowing, on the one hand (i), through the study of axioms, demonstrative principles and providing them to the other sciences (to dialectic and as a whole to all reasoning processes (*sullogizomenais*): that it is not possible to affirm and at the same time negate the same; to physics: that nothing comes from nothing; to geometry: that what are equal to the same are equal to each other; to all [humans] and

[sciences]:[13] that the good is what is the ultimate of desirables;[14] of such principles the sciences that use them will have a knowledge of the order of belief (*pisteutikê*), whereas he[15] will receive them from Intellect and know them in a way surpassing demonstration.[16] [And it also involves] him knowing, on the other hand (ii), through knowing the cause of being, all objects of science. Thus, just as it belongs to the same philosopher to possess science and the object of science, and in not reaching the one, he will lack the other, thus it will belong to the one and same man to examine the causes of beings and demonstrative principles, through which he will consider, from 3,1
above, from a certain intellectual vantage point (*periôpê*), both beings and knowledge of beings.

Third Problem

> **995b10-11** And if [this science] concerns substance, whether there is one science or several sciences dealing with all [substance].

It being agreed that wisdom concerns being, Aristotle wishes to know if wisdom itself will concern all being, or if several sciences will distribute between themselves the knowledge of being.

We answer that it belongs to wisdom to know all beings insofar as they are beings, but this does not prevent there being also other sciences which distribute between themselves portions (*moria*) of being, even if they study these in a different and in a more common way, sciences such as arithmetic, astronomy (*astrologia*), physics, medicine and other such sciences, as it may be.

> **995b12-13** And[17] if there are several [sciences], if they are all related, or if some of them are wisdoms, whereas others should be said to be something else.

This is a piece left from the preceding [problem]. And indeed Aristotle, in what follows, does not discuss it in particular.

But you might say, in response to the problem [*problêma*],[18] that the parts (*eidê*) of philosophy as a whole, such as first philosophy and physics, are related to each other and to the whole, whereas sciences which do not deal primarily with substances, such as the mathematical sciences, as not being ranked with wisdom itself or comparable to it in importance, beauty and value, would not at all be said to be related to it. Hence there are not many 'wisdoms', but only one, the truest. Yet the other sciences, as proceeding from it and as depending on the principles it introduces in them, not being able to detach themselves from it in any way, may be said, in this way, to be related to each other and to this science which is also most directive of them.

And [one might say] that Intellect is the transcending father of all sciences even if [the father most] immediately of wisdom, which he brings into being, a monad, as it were, of the sciences, which precontains in itself, as having their principles, all the forms of knowledge, a father who projects though [wisdom] the principles of the more particularised sciences. Thus the sciences are not all of the same value, but the science which is nearer wisdom is more worthy than one which is further removed from it, and a science which has to go though several other sciences to reach wisdom is inferior to all the intermediary sciences. Nor are the sciences related in the same way. For arithmetic is more related to geometry and the other mathematical sciences than to physics and medicine, for it is brought back with them to Intellect through one link, that of wisdom. But dividing, analytical, definitional and demonstrative [sciences][19] fit better with each other than with others, since they are closely dependent on wisdom itself and are given through it, with a very bright fire,[20] to the class of souls that are to be saved.

Fourth Problem

> **995b13-15** And this is one of the things it is necessary to discuss, whether it should be said that there are only perceptible substances, or whether there are other substances besides these.

Aristotle says that it is worth enquiring if there are also other, intelligible substances, as well as perceptible substances such as the heavens, the earth, the intermediate elements, animals and plants. But it would more appropriate to ask if, as well as true substances, it is right to speak of the world of appearances as substance (*ousia*), for perhaps a more appropriate name for the latter is that of 'coming-to-be' (*genesis*).[21]

> **995b15-18** And if there is one or several kinds of substances as is held by those who posit both Forms and mathematical objects intermediate between Forms and perceptible objects.

This also is a portion of the preceding problem, for the whole will be treated as one in the following. What Aristotle is saying comes to this. Even if it is agreed, he says, that there is a [kind of] substance besides perceptible substance, will it be of one kind, or at least of two kinds? For Plato seems already to have accepted, as intermediate between intelligible and perceptible substance, discursive (*dianoêtê*) substance under which he also included mathematical objects. By these objects is not meant what is grasped by imagination (*phantasta*) and by opinion (*doxasta*) – for these are not substances, but images of substances – but rather those objects that are included in the very essence (*kat' ousian*) of soul, objects of which the demiurgic Intellect, as the account of the making of soul in the *Timaeus* tells us, inserted the arithmetical, geometrical and harmonic principles in the soul.[22]

We will therefore say that it is possible even to speak of all intelligible substance as being one, when we separate only perceptible substance from it ('what is always, having no coming-to-be, and what becomes, never being?'),[23] according to the division in two of whatever in some way is. But it is also possible to subdivide invisible substance known by reason (*logismôi*) into what is properly intelligible substance and what is discursive substance, according to the [division of] the line in the *Republic*.[24] And since it is also possible to subdivide each of these (for we accept many ranks of intelligibles and intellectuals), the realm of souls too will show itself to include in many ways much differentiation with regard to substance, to whomever cares to distinguish the kinds of beings.

Fifth Problem

> **995b18-20** And this, therefore, as we say, we should examine, and also whether the study simply concerns substances, or also the essential accidents of substances.

Does wisdom, he says, just examine the substances of things, or essential accidents? We will say: both substances and what belongs in this way to them. For by the analytical [method] wisdom grasps the principles of being, by the divisional and definitional [method] the substances of all things, by the demonstrative inferring the essential properties of substances. This, however, is not the case with substances which are the most simple and properly speaking intelligible, for these substances are entirely that which they precisely are (*hoper eisi*). For this reason they cannot be defined or demonstrated, but are grasped only by apprehension (*epibolê*), as Aristotle often states, saying 'intellect either touched or not',[25] as does the divine Plato: 'only grasped by the governing intellect of soul'.[26] But it is the case for intermediate substances, which can be demonstrated as regards the properties in them.

This is the situation: for the simplest of beings, there is nothing belonging to them besides their being (*to einai*), so that for them there is not both substance and something else; therefore they are beyond both definition and demonstration. However, essential accidents do belong to universal rational principles (*katholou logoi*), both taken in themselves and as they order perceptible nature; thus demonstration concerns these accidents. And [these] accidents, properly speaking, also appear in forms that are enmattered, individual and by now perceptible: they come to be and disappear without [involving] the destruction of their substrate. Such accidents, in turn, falling short of what can be demonstrated, come to be grasped through a process of imaging (*eikotologia*), not indeed by the wise man as wise,[27] but, for example, by doctors, students of physics and all suchlike.

Sixth Problem

> **995b20-5** In addition, concerning the same, the different, the similar and dissimilar, contrariety, the prior and posterior and all the rest which dialecticians seek to examine basing the examination on received opinions only, who is to study all of this?

Now that it belongs to the wise man, according to Aristotle himself, to examine both substances and essential accidents, this is what the preface (*prooimion*) to book 4 proclaims, where it says 'There is a science which studies being as being and what belongs essentially to it' (1003a21-2). But as regards the matters now at hand for study, they would seem especially to belong essentially to being, as might be shown by their being found throughout all beings. For difference, similarity and dissimilarity, and such things as he now presents are not to be found as present only in some beings, and as absent from others. But starting from the intelligible [realities] above, ordering all that which is intelligible and divine, [going] through psychic substance, these come down to the nature of the universe (*phusis tôn holôn*), to the heavens and to what is in becoming: from the nature of the One which is absolutely good (*panagathos*) they impart to beings identity, equality and similarity; but they also confer both on invisible and on perceptible substances differences, dissimilarities, inequalities, contraries, the prior and posterior, and all such things, deriving them from the most productive and inexhaustible cause of all things, the infinitely powerful Dyad. Thus there is nothing that is not ordered by these forms, neither a nature intermediate between indivisible and divisible [beings], nor a [substance] transcending in this way in its excellence the whole (*tôn holôn*), nor one which has declined to the ultimate of perceptible effects (*dêmiourgêmata*). And it is for this reason that Plato, investigating in the *Sophist* the kinds of being which traverse all beings, enumerated identity, difference, being, stability, movement, and showed as well the nature of what is not, which holds the one cause both of all difference and of opposition and contrariety.[28] And furthermore, in enumerating the first of the Forms in the *Parmenides*, he mentioned similarity and the others, both prior to the hypotheses, in the arguments (*skemmasi*) about the Forms,[29] and in the hypotheses themselves: by means of the first hypothesis showing that these are transcended only by the goodness, beyond being and all multiplicity, of the One; by means of the second hypothesis[30] ordering all beings through these causes. And yet if these cannot be present in what is beyond being, nor are they absent from whatever in any way might be, how would one deny that they are essential accidents of beings as beings?

Thus those who specialise (*deinoi*) in nature, and the mathematician, and he who studies first philosophy must use these [essential accidents], the former on the basis of belief (*pisteutikôs*),[31] whereas he only who does first philosophy deals with them in a scientific (*epistatikôs*) and intellectual way. So it is that the demonic Aristotle himself,[32] both in book 10 of this treatise, teaches [us] in all these subjects, and here showed implicitly that it belongs to the wise man to know these [essential accidents]. For if the dialectician will deal with these on the level of opinion, as is said both here (995b23) and in the *Topics*,[33] who will provide scientific teaching concerning them? Is it not clear that it is he who is impersonated by the dialectician?[34] The latter does not pretend to be an expert (*tekhnitês*), for else he would not have begun to attempt to speak about everything, but only about those matters which come within the expertise he pretended to have. Now, however, desiring to appear as a wise man,[35] he has questions and answers about everything. Therefore the wise man speaks in a scientific way about all those matters which the sophist discusses on the level of opinion (*endoxôs*).[36] For he also is a distant image of the wise man, the dialectician being a nearer [image], just as opinion is nearer to intellect and the imaginative (*to phantastikon*) is further removed from it. If therefore these impersonate the wise man in dealing with these matters, he, *a fortiori*, will know their nature and what is proper to them, since neither would the images[37] be working with them, were they not to perceive the complete, primary and most true science as treating of them.

The wise man will therefore deal with these matters themselves and with what belongs to them, examining them as regards what is proper to them: if what are equal to the same are equal to each other, and if one is contrary to another, and suchlike. For everywhere the knowledge of what something is means knowing and demonstrating what belongs essentially [to it]. And indeed it would be absurd to possess the first kind of knowledge of whatever, but to be ignorant of the second kind of knowledge which comes after it. So much on this subject.

But if someone wants to know why Plato[38] includes among these [essential accidents] rest and movement, whereas Aristotle omits mention of them, he should know that Plato, knowing motion and rest <include>[39] the realm of divine and human matters, necessarily enumerates these also among the kinds of being, whereas Aristotle, believing that only natural bodies are in motion and rest, plausibly does not think that what is proper to these kinds extends to the totality of beings.

Seventh and Eighth Problems

> **995b27-31** And whether the principles and elements are genera or are that in each thing into which it is divided. And if genera, whether that which in individuals is said to be the last or the first, such as animal or man, is a principle and is more, besides the individual.

Two problems are presented by Aristotle here,[40] (i) if one should speak of the *genera* of things as being principles and elements, 7,1 principles as being causes, elements as being more simple (since indeed definitions are resolved into these last parts), or rather of those *components* in a thing into which the thing is divided, which are also said properly <to be the elements>[41] of it; and (ii) if one were to say that genera are especially principles, and not that into which each thing is divided [the components], whether these genera are those which are more comprehensive and wider in extension (for the concept [*ennoia*] of principle would suggest this), or those which are predicated more narrowly, for example, whether the principle of Callias is animal or man.

Let us say, then, in regard to the first problem, what kinds of principles are in question in the discussion here. For if the problem concerns material principles or principles belonging to the form, then what are present in things [components] are principles. But if efficient or final principles are at issue, then genera are principles. But these genera are not these,[42] for these derive from principles, <nor> what comes after (*husterogenê*), for the latter supervene on constituted things.[43] Nor are they that which are co-ordinate with perceptible things. For how could the man and animal *in* Callias be his efficient or final principles, since they are parts of the visible substance? But if there are certain genera prior to particulars, causes of perceptible things, which can be found both in the rational principles of the nature of the universe and far prior to this as shining in the forms belonging to the world-soul, these we would describe rightly as the causes of things here below. And prior to these <and above them are demiurgic forms which are causes and which>[44] transcend in simplicity universal rational principles. In these one should add to the efficient cause the final cause, not because [the latter] is not present also in lower principles, but because it manifests itself more clearly among the [higher] principles since these are established finally in the vestibule of the Good.[45]

As for the second problem, the answer is clear <from what>[46] has already been said. For if we take,[47] as genera, [i] those which come later by thinking (*kat' epinoian*),[48] those which are closer to what are said to be first substances[49] will be substance to a greater degree and, because of this, principles to a greater degree, whereas those which are further will be less substance than those[50] related to [first] substances, since they have a fading obscurity through [their] distance from perceptible things. But if we take [ii] what are, in the proper sense, universal genera and the rational principles that produce particulars, on the level of nature and of the demiurge, it is evident that the more comprehensive and more pervasive it is, the more divine a cause it will be and nothing anywhere will be able to proceed or act without it. It, however, makes use of what comes after itself whenever it produces something in relation to it,[51] whereas it brings to existence many others which do <not>[52] need a more partial cause.

Of these [matters] there are many clear indications given even by the image-like genera and species (*eidê*) used by the dialecticians. For of those [things] of which the species [is predicated], of these also the genus [is predicated], if it is said to be in the whole of it. But for many other things, of which the genus is predicated, the predication of the species is not such as to suffice. If therefore one switches from the predicating of certain things to the producing of them, one will move from images to true genera and species (*eidê*) and one will see the way they are co-ordinated with each other, the superiority and inferiority [among them] and in general the value of each of them.

Ninth Problem

> **995b31-4** And one should especially investigate and deal with the question whether or not there is a cause in itself besides matter, and whether or not it is separable, and whether it is one or many in number. 8,1

The natural philosophers, seeing only matter,[53] said it was water or air or fire, while others allowed also that there is an efficient cause, but saw it as not separable from matter, as in the case of the Stoics, later on, and of some [philosophers] before them. Others allowed that there was a cause separable from matter, as he [Aristotle] did and Plato too, although Aristotle posited this cause as being the object desired by all things, whereas Plato saw it also as generative of all things. So Aristotle reasonably presents [the problem] in this regard as to which of these opinions (*hupolêpseis*) is more true. It is clearly his own, as well as that of Plato. But when he then wants to know whether the separable cause is one or many, we will say that it is indeed both one and many, the many being co-ordinated with reference to the one and tending to it as what is appropriately desirable, just as he himself also tells us in book 12 (1075a11-25).

Tenth Problem

995b34-6 And whether there is something besides the whole (by the 'whole' I mean whenever something is predicated of matter), or not, or for some there are, for others not, and what kind of beings these might be.

In these [words] he enquires if, besides the composite formed thing, there are universal and essential Forms (*eidê*), and if they are Forms of all beings, both natural and produced by art, both beautiful and ugly, good and evil, perfect and imperfect, or if there are Forms of some, but not of others, and of which in either case.

Now whoever wishes will find an accurate articulation of these matters in the discussions in Plotinus[54] and Iamblichus[55] concerning the Forms. And we will develop an investigation elsewhere[56] into the question whether there are also Forms of individuals that are in becoming, or of parts (such as the foot or the finger), or of qualities, or generally if there are Forms of accidents in nature, in the soul, or in intellect. But let us say here something about divine and intelligible causes, more as an outline and as much as is necessary.

We say then that there are no Forms (*ideas*) at all of what is ugly, imperfect and evil. For these arise as a result of a declension (*apoptôsis*), on the level of the lowest beings, of nature, or of particular soul weakening due to not mastering the underlying indefiniteness. There are however demiurgic Forms of those substances that have been constituted naturally and exist always, just as there are forms of artefacts in the skill [that produces them]. Now Aristotle even agrees that there are forms of what is produced by art, for he says in many places that the enmattered house is produced from the immaterial house, just as is said in book 7 of this treatise (1032b12). However, he does not furthermore allow, as part of what is distinguished as form, the paradigm of the arts, I mean the whole production (*dêmiourgia*) [of the universe]. Yet how could that which imitates nature [i.e. art] alone produce in this way, if it were not the case, far prior to it, that nature herself produced in this way?

Eleventh Problem

996a1-2 Then, if principles are determinate in number or kind, both those in statements and those in a substrate.

It is taken here as agreed that principles are determinate. The 9,1 question is (i) if the principles are determinate in number, as for example the four elements of bodies (for these are four in number, and so body does not simply come from air and fire, but from cosmic air and the fire here), or (ii) if they are determinate rather in kind (*eidei*), like the twenty-four elements[57] of speech (*phônê*). For it is clear that these are determinate in kind, but not in number, from the fact that 'a' as a whole is present in many syllables, not a part of it in 'ba' and [another] part in 'ga' in the way that a part of air constitutes this body, and [another] part another body. So he wants to know if principles are determinate in number or in kind.

It is clear that if they are determinate in kind, it will not in every case also be in number, but if they are determinate in number, then also in kind. His statement 'both those in statements (*logois*) and those in a substrate' is due either to the difference among principles, so that he can say that formal, efficient and final causes are principles in reason-principles,[58] but material causes are principles in a substrate; or he is saying that the principles that are proposed on the level of opinion (*endoxôs*) are in statements, whereas those which are truly present in substances are in a substrate.

Such being the subject of enquiry, we will say that both of the true principles and causes of all things, the One and the Indefinite Dyad, transcend that which is included in [Aristotle's] distinction.[59] For [i] the individual and one in number fits only with enmattered forms, since the immaterial one is not less all than one. And [ii] as for what is determinate in kind, if it is considered on the level of elemental principles,[60] it carries an image of a superior unity, as do the many (or [rather] there is in every case an image of both in an affection of reasoning or a signifying representation (*phantasma*)), for it is in this way that the many stones and horses are images of a unity. But if it [the determinate in kind] is efficient and generative and transcends what is produced, or is desired by them and perfects them, it comes directly *after* the principles of the whole (for the world of Forms comes below[61] the principles and is the first [derived] from them), but is not what precisely are the principles of the whole, which even transcend in simplicity intellectual and form-numbers.[62]

However the principles which Aristotle himself teaches in book 12[63] would seem to encompass both.[64] For indeed these are both determinate in number, if indeed they are as many as the spheres of circular movement, and determinate in kind, if indeed each of them is different in kind from whatever else there be among them, such that one is higher and makes use of a more comprehensive and simpler thinking, whereas another is inferior and subordinate, being ordered to the body-like (*sômatoeidesi*) spheres in the same relation.

Thus, on the one hand [i], if we consider *Pythagorean* principles, neither of these [the determinate in number and kind] are among the principles. For how would it be at all possible ever to find in the principles a determinate multiplicity of any sort, if all multiplicity, determinate in relation to the individual one or in relation to kind, is number, and all number is a product of principles? But if we were to say that there is one principle of all things and if we were to refer to it as God, or the Good, or One, <or if we were to speak of the first

principles as>[65] limit and the unlimited (as Plato in the *Philebus* and Philolaus before him name these principles), or as the monad and dyad (as do the majority of the Pythagoreans), or as aither and chaos (as does Orpheus), or as Prateus[66] and dyad (as does Pythagoras himself in the *Sacred Discourse*),[67] we do not properly name them (for they are not only above all naming, but also above all human thinking), nor do we name them the One and Two according to number or kind (for the principles precontain in themselves the cause of all number and all kind, intelligible and intellectual), but there is another way of indicating these principles to be found in those who [have ascended] higher in divine science (*theologikôterois*).[68] So much then for the Pythagorean principles, which is to say those of the Orphics and the Platonists.

But [ii] as for *Aristotle's* principles, on the other hand, let us say that they are determinate in number and kind. For they are indeed so and so many, and not altogether similar in kind, but differ by superiority of substance and by value. But what is different as a particular (*aphorismenôs*) does not differ by reason of what are properly speaking principles, but if at all, by reason of material and formal principles.[69]

Twelfth Problem

> **996a2-4** And whether there are the same or different principles for perishables and imperishables, and whether they are all imperishable, or the principles of perishables are perishable.

He seems to present these [matters] (<if there are>[70] the same principles of perishables and imperishables) as if there were two problems: for if they are the same, how do they produce some as imperishable and others as perishable?; and if they are different, whether the principles are all imperishable, or some are imperishable and others perishable. However one might like to distinguish these things, we reply to him that the principles of what is imperishable are the principles of everything that is, in one way or another. For their generative and infinitely powerful activity is nowhere circumscribed. But the principles of what is perishable are not all principles also of what is imperishable. For instance the proximate principles of perishables are the principles of them only. However neither are all of these principles themselves perishable. For the circular motion [of the heavens] has, according to Aristotle, the efficient cause, but it is not perishable. And there are other imperishable causes of things that are in becoming, as indeed there are also perishable causes. For in general if man is generated by man and the sun,[71] it is evident that man has a cause that is perishable and one that is imperishable. It is the same also for horses, dogs and for whatever else is part of the things here below. For each of these has both a partial and a general cause. But it is absolutely impossible that the more general principle be destroyed. For otherwise this cause would not be generated from something, nor would other things be generated besides it.

So it has seemed, rightly, not only to the divine Plato, but also to Aristotle. For he says that the principles of the eternals, such as the separately existing immaterial forms,[72] are desired by all, and some of the principles of perishables are eternal. At any rate, he will often worthily prove (also in what follows) that there would be no becoming if there were not some eternal cause for it. Yet it is to this extent that he falls short of the philosophy of his father,[73] when he does not assign efficient and paradigmatic causality to immaterial forms, but a cause that is final and an object of desire. For, according to him, these are desired proximately by the spheres that circle around them, and through these, they are desired by all things in the cosmos. For he also says that all things desire the good, and if there are many goods, they are all ordered to those which are superior and these to one, the highest and most perfect of all, as he clearly says, we can see, in [book] 12.[74]

Thirteenth Problem

> **996a4-9** And then there is the hardest of all, which brings the greatest difficulty, whether unity and being, as the Pythagoreans and Plato said, are nothing other than the substance of beings. Or not [so], but the substrate is something else, as Empedocles says in regard to Love, and as another [says] in regard to fire, and another to water or air.

One might well admire the fair-mindedness (*epieikeia*) with which Aristotle does not think that the opinions of elders are lightly to be despised, but [require] much attention, especially the opinions regarding the very first principles. And I think he will agree that it is necessary that, there being many visible and intelligible substances, all of them depend on one principle, which one might characterise as being that which primarily is. But what he does not say from this point on, but which necessarily follows from what he posits, this it is for us to say. And so we say that all beings would not desire that which primarily is, were it not the case that they acquired their perfection from it; and that that on which they depend for all eternity, from this they also received eternally their being. Consequently, if that which primarily is is desired by all beings, and it is the cause of being for all, [then] it is nothing other than that which primarily is, so that it may be the cause of all beings, producing from itself substantial number and what are truly beings and intelligible forms.

But since Being is this, and, if it is the principle of all beings, it is in some way ranked with them and because of this not entirely without a hint of multiplicity, one must consider, as prior to it, the unity that is beyond being, transcending all beings in its simplicity and inconceivable excellences, which it is impossible to name properly. But one might speak of it as the One more appropriately than in any other way, because it is the cause of unity for all beings, which causes assimilation to it.

So it is that the Pythagoreans necessarily posited unity and being as prior to all things, the former as the cause of unity and of all goods for beings, the latter as providing the primary principle of being also to all the other forms. And I do not think that Empedocles posits, with Love, anything other than the One, although it is not the One that is not co-ordinate with all things, but it ranks with the Indefinite Dyad, which he calls Strife, from both of which arise what primarily is and all intelligibles and the perceptible world-order. For if, according to this philosopher, Love is the cause of unity for the sphere, which we relate to what primarily is, and Strife is the cause of multiplication and otherness and generative progressions, why would Love not be understood as the One and Strife as the Dyad? Since also Empedocles is a Pythagorean, how could he have rejected Orphic or Pythagorean principles? But if Thales said that unity and being are water, and another philosopher said that they were one or
12,1 the other things that appear, their opinions have been evaluated by many and most especially by this demonic[75] man [Aristotle].

Fourteenth, Fifteenth and Sixteenth Problems

> **996a9-11** And whether the principles are universal, or as the particulars among things, and [if] in potency or in act, and again if in a way other than that according to movement.

The principles, properly speaking, are known as both above universal genera and forms and even more so as above particulars. For it is like this. Particulars are seen in matter, but in nature and in soul universals pre-exist as the causes of perceptible things: nature containing as it were the most specific forms from which are immediately (*prosekhôs*) generated enmattered things and particulars; soul containing both these [forms] and, prior to them, more universal rational principles (*logoi*) through which soul, in dividing genera and then again defining the multitude of rational principles, knows all, descending and ascending and in general acting by division, analysis and definition. Above these are placed demiurgic forms, and then again, above these, superior to all, are the principles. This much as regards the first question, if large matters are to be said in few words.

As for the second question, we must exclude in every respect what is in potency from principles in the proper sense. For what is in potency is imperfect and non-generative and more appropriate for matter. But what is in act is more appropriate to the principles, except that some of these are said not to be in act, but to be this alone, act.[76] However the Principle of all things, if one may say, would not only be above what is in potency and in act, but also above act itself.

As for the third question, we say that the causes of eternal things are unmoved principles, whereas the principles of things subject to generation and destruction are moving, principles such as the sun and the circular movement of aither.

Seventeenth Problem

> **996a12-15** In addition to these questions, whether or not numbers, lengths, figures and points are substances, and if they are substances, if they are separate from perceptible things or are in them.

It would seem that, in these [words], 'substances' (*ousiai*) is understood as 'beings' (*onta*), for he did not think, I presume, that the point in perceptible things and the line are substance, though nonetheless he will investigate in what follows what is substantial in them. We, however, show both their being and what is substantial in them in saying that they are of many sorts. For indeed one might see in the perceptible works of nature both figure and number and natural surface and its limits. Furthermore, these are constituted also in our representation (*phantasia*) and in opinion (*doxêi*), either by being taken, as Aristotle thinks, by abstraction from perceptibles, or by being produced in us from the substantial forms of the soul. Now the objects of representation and opinion share in being, but are not substances. Thus one might assign them rather to quantity, to quality or to another category. But the substantial principles (*ousiôdeis logoi*) of soul which contain these[77] are already substances. And if one were to consider the paradigms of them in Intellect and in 13,1
the intelligibles, one would see that number and figures and magnitude itself are numbered among the very first of substances.

Conclusion of the First Part

Having presented all of these problems, our philosopher [Aristotle], starting up again from the first problem, attempts an exercise (*gumnazein*) in arguments that oppose each other, in order that, hearing the opposing arguments, we may ally ourselves with what is more true through much testing and attentive discrimination. Now, while indicating[78] that Aristotle does not everywhere follow the order of the problems, let us now try to accompany him, as he exercises with the opposing arguments, agreeing with what points to the truth, resisting what is more sophistical. For it is necessary that of his opposing arguments, some look to the truth, some to deception.

Chapter 2
Second Part
Discussion of the Problems

First Problem[79]

> **996a18-20** First, then, of the problems of which we have spoken: whether it belongs to one or to several sciences to consider all of the genera of causes.

This was the first of the problems he presented. By the 'genera' of causes he means here the kinds. Then [A][80] he will first argue that it does not belong to one science to know all the causes of the object, which is both false and not his opinion.

> **996a20-1** For if it belongs to one science, how could principles be known which are not contrary?

The syllogism would mean something like the following. Causes which are different in kind are not contraries; there is not one science of those [things] which differ in kind and are not contraries; so there is not one science of causes. And (an argument in the second figure):[81] causes differing in kind are not contraries; those [things] which come under one science, [even] if they differ in kind, are contraries; causes do not come under one science. In whatever way, the major premise[82] is in every sense completely false. For it does not follow, if there is one science of contraries, that there is not one science of what are not contraries. Nor does it follow immediately that if contraries come under one science, what come under one science are contraries. But perhaps causes do not always differ in kind, as Aristotle notes in book 2 of the *Physics* (198a24-7), that sometimes final and efficient causes coincide with formal causes in relation to what is similar in kind.

> **996a23** Then, for many beings, there are not all the principles. For in what way can there be, for unmoved beings, a principle of movement, or the nature of the good?

The syllogism runs like this. The primary and finest sciences deal with unmoved beings; sciences dealing with the unmoved do not know all the causes, since there is no efficient or final cause in what is unmoved; [thus] the primary and finest sciences do not know all causes. Here again, the major premise is false. For let the theoretical
14,1 sciences be the primary and finest of sciences, and of these let

physics, which deals with moved things, not be preferred, but the more mathematical sciences and first philosophy be more valuable than it. Why then do these sciences not consider all of the causes of the objects of science? 'Because', some might say, 'unmoved beings do not have an efficient or final cause'. On the contrary: they especially have these, if indeed they come from a principle and have not turned up (*parêlthen*) spontaneously among beings. For all that which in some way is, is either a principle or from a principle. But a multiplicity cannot be a principle, so that it should have an efficient and final cause. For it is on account of the good that unmoved beings [are] knowing and are known.

Then again, the initial subject under discussion is different. For what is to be examined is not if there can be a science of the object if a cause is missing, but if there be, not one science, but many which study the same things, if there are different causes, a point not argued here.

Rather, one might, starting from these matters, reach the opposite conclusion. For if all the sciences which deal with unmoved beings know all the causes which are in the subjects [of study] and overlook nothing which is in them, wisdom[83] too would know all the causes of being as being.

The following argument (996a29)[84] tries to prove that mathematical objects do not include a final cause. But it is falsely taken that the 'that for the sake of which' is the finality of an *action* (*praxis*). For there are final causes, not only for actions, but also, and much more so, for [forms of] knowing.[85] For we desire to know for the sake of the good and assimilation to the divine[86] and our true salvation, whereas the finality of action is both partial and not finality in the proper sense.[87] And it is not to be wondered that geometers do not use statements of the final cause when they show the essential properties of figures. For neither do they examine their subject-matter, if it exists or not, nor do they busy themselves with the final cause nor with anything else but the essential properties of figures and magnitudes. And it is for this reason that they would not have scientific knowledge.[88] 'For him who does not know the principle', he says,[89] 'and does not know the end and the intermediates out of which [it is made], how could such an agreement ever become science?' But he who does first philosophy fully examines the reason why the circle inclines inwards to itself, and what it imitates of those beings that really are, and what is the first of the figures, and what is their order and whence: [all questions] in relation to which the causality of the Good is wholly involved.[90] And Aristippus absurdly thinks[91] that the culinary art prepares this [kind of] food on account of some finality, but does not think that the mathematical sciences, which always reach true conclusions from scientific premises, show the things with which they deal as the products and imitations of Intellect. For whence comes the truth in them, whence their order or unspeakable beauty,[92] whence the benefit they bring to souls in turning them around and leading them up, were it not far more that there was given them from Intellect a large share of the Good?

> **996b1-6** But if however there are several sciences of causes and a different science for a different cause, which of these should we say is the science we seek, or, among those who have them, who especially knows the thing sought? For it is possible for there to be in the same all the types of causes.

[B] He proves [here] the opposite, which is indeed true: that one science knows all the causes of the object of knowledge. And this is his own opinion. The proof of the whole position is strong, concise and clear. But some intervening steps require brief examination. He says that if one science were to know the efficient cause, and another the final cause (for the thing can have all causes), there will be both many wisdoms and they will contend for primacy and in truth none of them will be wisdom. For they will all be imperfect, since they have distributed among themselves the knowledge of the causes. For a science would be ridiculous, claiming as it does to know the efficient cause, when it neither knows what the thing is, nor for what reason it came to be or is. And a science would be imperfect which says that it examines what something is, if it knows neither the principle of movement, nor the 'that for the sake of which'. And a science which possesses, so it thinks, the final cause, but knows neither that of which it is the final cause, nor whence the thing came to be, how is one to take it seriously? So the science is one which considers all causes. These points contain the full force of the proof.

Let us look briefly at the phrase:

> **996b6-8** As in the case of a house, the source of movement is the art and the builder; 'for the sake of which' the function (*ergon*); the matter is earth and stones; the form the plan (*logos*).

These words prove that all causes are present in some things. He says that the 'for the sake of which' (the final cause) of the house is its function, i.e. to shelter us and protect the necessary equipment. For this is the function of a house and it is because of this that it first came about.

> **996b8-10** Now from what has been determined before about which of the sciences is to be called wisdom, there is reason for each to be named thus.

Concerning wisdom it has been said in book 1[93] that it is more architectonic than the other sciences, more teachable, knows most and is most accurate. He says therefore that, in accordance with these specifications, if there is not one science that knows all causes, but each science knows one cause, the sciences will dispute with each other. For the best and most ruling science would be that dealing with the final cause and the good, whereas the most demonstrative and most definitory science would be that dealing with the formal cause. But in actions and in all moving things, if one does not know the source of movement, one knows almost nothing of the thing. He opposes the efficient cause to the formal cause, since the one is the cause of movement, and the other is the goal. We should attend again to a particularity of his philosophy: that he accepted an efficient cause for moving things only, but none for unmoved things.

These points are to be gathered from a number of places,[94] but let us hear how he says that the science that knows the final cause rules (*basilikon*) :

> **996b10-13** The science which rules and leads most, which it is not right for the other sciences, as its slaves, to contradict, the science of the final cause and of the good is such[95] (for the others are for the sake of this).

For inasmuch as wisdom possesses what rules and what dominates the other sciences, it would be none other than the science of the *final* 16,1
cause, for the sake of which the others are. And indeed the final cause has a ruling value in relation to the other causes. For to that for the sake of which everything comes about, all is subject, in the universe, in the state (*politeia*), in an army, at home and in each of us: all come about on account of the good and the final cause.

> **996b13-18** But the science which is determined as being the science of the first causes and of that which is most knowable, this, the science of substance, would be such.[96] For of the many ways in which things are known, we say that he knows especially who knows what a thing is by [its] being, rather than by its not being, and of these the one more than the other, and most especially he [who knows] the 'what it is', and not the quantity or quality or what it is of a nature to do or to suffer.

He says that inasmuch as wisdom is of the first causes and of that which is most knowable, wisdom would be the knowledge of *form*. He will go on immediately to say how these belong to the science of form, but first he shows in this passage how, among the many who have knowledge of things, he [who knows the form] has better knowledge. For if he also somehow knows who can speak of something by means of privation or negation, for example he who knows that the point is indivisible and the divine immortal, he has greater knowledge than the former who knows it through affirmation and its properties, for example that the point is the limit of a line, or that the divine is the eternal self-sufficient good and cause of beings. And then again, he has better knowledge [than the former] who, rather than being able [merely] to speak of things in relation to their quantity, quality, how they act or suffer (for all this comes after substance and is secondary to it), can teach about the 'what it is' of [his] subject: he, especially, would be the most perfect of those who have knowledge. And as for how this relates to first causes, let us hear Aristotle's own argument.

> **996b18-22** Furthermore, as in other cases, we especially think we know each thing (and those of which there are demonstrations) when we know what something is, for example what it is to produce the square of something, that it is by finding the mean, and likewise for the rest.

Now substances are known through definitions (thus first science uses [the method of] definition), whereas essential properties are known through demonstration (thus secondary science is demonstrative). But the science that has examined form, knowing all of these, is both definitory and demonstrative: definitory in an immediate way (*autothen*), since this is what it would be to know the substance of a thing, demonstrative, since in giving causes, one will not stop the demonstration before reaching the 'what it is'. For it will not suffice for us[97] to say that fire separates, whereas snow joins, for example, but we should say why each is so. 'The one', he says,[98] 'is hot, the other cold.' But why is each of them so? If we do not reach as far as the definitions of these things and show that this is what it is for each to be fire or snow, we do not have demonstrative knowledge. Thus only he who knows the form can have demonstrative knowledge. And this [the form] is the first cause, which is taken from the definition, 'first' not in the account given (for in an account it may come last), but in the nature of things, where, when it is reached, demonstration ends. For example: 'why does the statesman (*politikos*) want there to be music and gymnastics in the city ? So as to educate the citizens. Why educate? To make them knowledgeable. But why this? Because they are receptive of knowledge. How can this be done? Because they are rational. Why are they rational? Because this is what it is for them to be human.'[99] You see that he [who knows here] knows from first causes and that the best cause in a demonstration is the definition of the thing. 'Why is there an eclipse of light? Because light is cut off by the earth.' You thus have the definition of a lunar eclipse. For a lunar eclipse is the privation of the light of the moon due to obstruction by the earth.[100] If you wish to demonstrate this, you need merely to exchange the position of the terms, as Aristotle has shown us in the

Demonstrations.[101] And the squaring of a right-angled parallelogram is nothing other than the discovery of the mean term.[102] For if its length, for example, is 9, its width 4, the surface area will be 36. Now, in finding 6 as the proportional mean of 9 and 4, you draw with it a rectangle equal to the parallelogram. But whether you would say what it is to make the square of something (and you will say it is nothing other than finding the mean proportional), or if you would demonstrate that it has been squared, you will in any case end up with this, the very first cause, and with no other. Thus, only the science that knows the form both defines, and, as for the rest, i.e. essential properties, only it demonstrates them.

> **996b22-4** But as for coming-to-be and actions and all change, [we think we know them] when we know the cause of movement; this is different and opposed to the final cause.

It is shown by this that the knowledge of the *efficient cause* is most necessary when it comes to actions and in general to things which move. For he who does not know the principle of movement will not know the form or the finality of the thing. For instance we have often been satisfied to know that the action is that of Socrates or Pythagoras, in order to be persuaded that it is fine, wise and beneficial.

> **996b24-6** So that it would seem to belong to another science to know each of these causes.

Either this is written in such a way that we will understand Aristotle as leading us to an absurd conclusion, as if he had said: 'so it would belong to one wisdom and then to another to know each of the causes,' which is to say that there will be many wisdoms, if one wisdom does not know all causes. Or '*not* to belong to another [science]' [is written], as is rather Alexander's judgement,[103] so that we should understand Aristotle as arguing directly from something which is shared,[104] as if he were saying: 'since wisdom knows the cause, it does not belong to different sciences to know the causes, but to one'.

Second Problem

> **996b26-7** But indeed as regards demonstrative principles, it is a matter of dispute whether this [subject] belongs to one or to several sciences.

Having raised problems about the causes of being, if it belongs to one science to know all of them, or to different sciences, Aristotle, advanc- 18,1
ing meanwhile according to the presentation of the problems,[105] now

investigates the second problem, that as to whether it belongs to the same science to know both the principles of being and the principles of demonstration, or if it belongs to one science to know the former and to another to know the latter. And he will argue for both positions. But we have said before that it belongs to the one science to know all, and we will now try, following our original intention (*prohairesis*), to produce a refutation of the arguments for the opposed position.

996b27-33 By demonstrative [principles] I mean common opinions, on the basis of which everyone demonstrates, such as: that it is necessary for everything to be affirmed or denied, and it is impossible at the same time to be and not to be, and all other such premises, whether there is one science both of these and of substance, or another science, and if not one, which must be called the science that is now sought.

In these [words] he both explains what demonstrative principles are and defines the problem more clearly, subdividing each of the opposed positions so that the subdivision brings out another problem. He therefore shows that demonstrative principles are common notions (*koinai ennoiai*).[106] By this he does not however mean [common notions] inasmuch as they condition us or move us to action (for all things desire their finality in accord with a common notion, failing which [finality], humans, not able of themselves to possess it, succumb to having recourse to the divine),[107] and being moved by them, [humans] desire by nature, as he himself says, to know:[108] [acting in] this way, therefore, these principles would not be demonstrative principles. But [Aristotle means them] to the extent that we use them for demonstrative arguments. We call these 'axioms' because they are accepted by all in this way, and they constitute a knowledge which is more clear than that which is demonstrated. One of these [common notions] is that concerning contradiction. Our elders had two positions concerning this principle:[109] on the one hand, nothing escapes this principle, but all must necessarily be shown affirmatively or negatively, such that both premises cannot miss the mark in anything, but one or the other is true; and, on the other hand, that it is impossible for both to be true at the same time.

Aristotle accepts both positions here, but we will simply add to this that the second position is true absolutely, but that the first is true if taken with a prior specification. For it is necessary that everything either be affirmed or denied, when it comes to beings which can be grasped by science. Since if there be something which is above being and which has no name nor science and is wholly ineffable,[110] how could it be necessary for this to to be subject to affirmation or negation, when all discourse is false in relation to it? But these things belong to another, deeper discussion.[111] And indeed there are also further objections concerning this axiom for those who would not use it without intelligence.

When Aristotle says here: 'and all other such premises', we understand as implied in this phrase the principles that say: that all that is desired is good; that nature does nothing in vain, and still less the divinity; that no substance comes from non-being, nor is destroyed into absolute non-being; that all that which is moved, is moved in relation to something unmoved. Add to these the axioms of the geometers, of the arithmeticians and of the other sciences, concerning all of which no one other than he who knows the causes will investigate. For were it to be someone else – for this is Aristotle's difficulty – one must still explore which of both [of these people] would then be the wise man whom we seek.

> **996b33-997a2** Now it is not plausible that it should belong to one science. For what specifically would belong to geometry or to whatever science concerning knowledge of these matters? If then it belongs in like manner to any one science, and it cannot belong to all, so it is not specific to the others, nor to the science knowing substances, to know these principles.

He starts [A] by proving the opposed position, that the same science will not know substances and axioms, a position which is false. Let us see therefore where the syllogism misleads us. 'For if it belongs in 19,1
like manner', he says. 'to all sciences to deal with demonstrative principles, but the others do not know them, then neither will the science which deals with being know them'. Now the major premise (*sunêmmenon*)[112] is sound. But let us look at the minor premiss (*proslêpsis*): 'But yet it belongs to all sciences to deal with demonstrative principles in like manner'[113] [... both those who have] the principles of demonstration only on the basis of belief, and the wisdom[114] that receives them proximately from Intellect, knowing them from above in a mode superior to demonstration, seeing their truth together with their order and progression from Intellect. I speak here of axioms taken absolutely, since when axioms are taken in relation to a particular skill, and are not absolutely primary in the order of things (*phusis*), he who knows substance will also demonstrate these, although they are not demonstrable by him who practices a particular skill. For instance, he [the metaphysician] even demonstrates geometrical principles. How then could he stand in relation to these principles 'in like manner' as the others? Or how, if this is not conceded, could the error ever come about, the error to which the conclusion tends?

997a2-11 And at the same time, how will there be a science of them? For we know now what each of these principles might be (for they are even used as known by other arts). But if there is a demonstrative science of them, there will need to be some genus as a subject-matter, and some of them will be affects, and others axioms (for it is impossible that there be demonstration of all), for it is necessary that there be a demonstration from some things and in relation to something and of some things, so that it will come about that there will be one genus for all that is demonstrated, for all demonstrative sciences use axioms.

The argument attempted before to prove that no one science knows demonstrative principles, but it now tries to show also that these principles are not even objects of knowledge. Let us therefore see in what way this syllogism too tries to deceive. He says that the principles of demonstration are not known, unless it be to the extent that we know whatever they show. But this does not make a[115] science of them. But if they are known, then they are demonstrable, but if they are demonstrable, they have one subject-matter, just as numbers are demonstrable in having one subject-matter, of which some are affects (such as even, odd and the combinations, multiples and divisions of these), which indeed are demonstrated by arithmeticians, and some are axioms, as being the definitions of these. For it is not possible to have a demonstration of everything, as is shown in the *Posterior Analytics* (84a33). But if this [argument] imposes itself, then all demonstrative sciences will have to do with the one subject-matter of axioms, which is absurd. For as arithmetic deals with all numbers and geometry with all figures, thus each of the demonstrative sciences would deal with the whole subject-matter of axioms.

Now he took from 'if they are knowable', that 'they are demonstrable', which is false. And, from 'if they are demonstrable', that they have altogether a subject-matter, which itself is also false. For it is in one way that *things* are said to be demonstrable, and in another way that *arguments* and premises are such. But if axioms will have a subject-matter, affects and axioms will appear in relation to them, so that there will be axioms of axioms. He does not himself posit this absurdity, but it follows entirely from what [he] has supposed. And the consequence of this, that axioms, if they depend on other axioms, are not axioms, does not prevent our also going with this to infinity. For those axioms again, just as these, require other axioms, and so to infinity. Why then should we wonder, if such absurdities are agreed, that all sciences deal with one subject-matter, which is that of axioms?

But it is the case that axioms properly speaking are not demonstrable, and if they are objects of science, it is in relation to primary science which deals with unmediated premises which include the 'why' (*to dioti*) in the 'that' (*to hoti*),[116] since they proceed immediately from Intellect and explicate Intellect in dividing and using discursive [methods]. Nor would the axioms, if they were demonstrable, have one subject-matter, since they relate to all things[117] and are elevated.

[B] But it remains [now] for us to see how Aristotle proves what is true.

> **997a11-15** But if the science of substance is different from the science of these [axioms], which of these is more important (*kuriôtera*) and prior? For axioms are especially universal and the principles of all: if it does not belong to the [first] philosopher, then to whom else will it belong to consider what is true and false in regard to them?

If it is not the same person who knows the principles of things and those of demonstrations, but one person [the former] and another [the latter], then each of them will be deficient, the one not knowing how to demonstrate, the other spinning empty syllogisms since he is ignorant of the nature of things. But they will quarrel about who has pride of place, for there is also great prestige for him who knows[118] the demonstrative principles which include everything. And we will be in great difficulty as to how to name this man, if he is not identical with the wise man.[119]

Now he who remembers these things[120] will know that Aristotle has used the same method to prove that both he who knows all the causes of being [and he who knows] the demonstrative principles is the wise man, who is one. But it is possible to see that Aristotle has adequately shown the truth in a few steps. For if the division [in two] of forms of knowledge makes each science [of the two] deficient and ridiculous, pretending they know what they do not know, fighting furthermore over who has pride of place; and if the division shows each of these which attempts to encompass everything with its own intuitions (*epibolai*) to be of little worth and nameless, is not the union of these sciences most securely settled? And if one [or other] of these sciences provides demonstrative principles to all of the other sciences, what other science would this be but first philosophy, the science which deals with being as being? For first philosophy imitates Intellect, in which intellection and the intelligible are not divided, being assimilated to intellection through the knowledge of axioms, and to the intelligible through knowing being. But if it is deficient in either respect, it will be an imitation of Intellect as having one or other [part] and not both, as intellection without the intelligible,[121] or as the intelligible without intellection, which would be the same as imitating an unintelligible intellect!

Third Problem

997a15-16 And in general if there is one science or several sciences of all substances.

Aristotle enquires if there will be one science of all substance (intelligible [substance], perceptible [substance] and if there be a [substance] between these), or if there are several sciences which are differentiated according to the differences between substances. This was also presented as the third difficulty in the [series of] problems.[122]

Now a concise and completely true answer concerning these matters, with which Aristotle himself <agrees>,[123] says that there is one supreme science, and many proximate to it; one science that considers all being as being, and many which have distributed among themselves the parts of being, dealing with these in another mode and not as beings. But let us see how Aristotle now takes up the opposing positions.

997a16-17 Now if there is not one science, of which substance is this science to be taken as being?

[A] He says that if someone were to posit that there is one science for these substances and another for other ones, he will be puzzled as to which substance is dealt with by the science called wisdom. Now we will say that it is rather one science that will consider all substances; but if different sciences consider different substances, it is evident that the science dealing with the first intelligible substances will be the first of the sciences. Of course, even here, although it deals with all being as being, it is especially and most eminently concerned with the intelligible.

997a17-18 But that there is one science of all substances is not plausible.

[B] He moves on to show a difficulty which comes up also with the opposed position. But let us see what is said, or why it is not plausible.

997a18-19 For the one science would demonstrate all accidents.

And what is absurd in this? For even if there are many demonstrative sciences, such as geometry, astronomy, arithmetic, and many others, yet prior to all of these is one science which considers the essential accident of being as being. (What such accidents are, Aristotle himself will say most clearly in book 4).[124] For arithmetic shows the essential properties of number, and geometry those of figures. But the one demonstrative science shows the essential properties of all being, not as this quality or quantity, but as being. Thus nothing absurd follows. But he will try to argue that such absurdity does follow.

> **997a19-25** If indeed all demonstrative science considers, regarding one subject-matter, the essential accidents, starting from common opinions. Thus as regards the same genus, it belongs to the same science to consider essential accidents, starting from the same opinions. Concerning 'the what'[125] and 'from which': each belong to one science, if it is the same [science] or another, such that these sciences will[126] consider the accidents, or one from these sciences.

The argument is this. Were there to be one science of all substances, there will be one science of all essential accidents belonging to all substances, a demonstrative science, whether it is the same science as that which deals with the substances themselves, or another (for this is not yet the point at issue). But I show, he says, that there will be one demonstrative science. This is the demonstration. With regard to one subject-matter for which one science considers what something is, one science proceeds demonstratively concerning this, showing, starting from common notions, the essential properties in this subject-matter, these notions being the principles of all demonstration. For it is not the case that one science considers the 'what it is' of a subject-matter, i.e. its substance, whereas *many* sciences consider 'that from which', i.e. the essential properties, from which derive the definitions and what the thing is, but it is *one* demonstrative science, whether it is the same as that which concerns the 'what it is', or another, except that since the other is one, this itself is one. If then one science will consider all substances as one underlying genus, one demonstrative science will include also all the essential properties belonging to these substances, being either the same as wisdom or different. For this will be analysed in the next problem.

The phrase 'Such that these sciences will consider the accidents, or one from these sciences' means: so that, as for essential properties, if there are many sciences that know substances, all demonstrative sciences show them; but if there be one science, this one among the demonstrative sciences will consider all [the properties]. And this follows absolutely, even if indeed wisdom, which transcends all of the other sciences, is not one of the demonstrative sciences. But there is nothing absurd in there being one science which considers the essential properties of all being as being.

Fifth Problem

997a25-30 Furthermore, whether this study concerns substances only, or also their accidents? By this I mean, for example, if a solid body, lines and surfaces are substances, whether it belongs to the same science to know these, as well as the accidents relating to each genus concerned in what is demonstrated by the mathematical sciences, or to another.

Now this is the fifth in the presentation of the problems.[127] Yet I do not think that it is for no good reason that it is taken now after the third problem. But since Aristotle had posited this very problem in a very ambiguous way in the third problem ('for', he says,' either [it belongs] to the same [science] or to another'), he wished immediately to give priority to a discussion of this. Now enough has been said in the [presentation of the] problems about this, that the science that knows the subject-matter fully also demonstrates the essential properties, nor has what he now brings up any validity in refuting this point.

He says [A]:

997a30-2 For if it belongs to the same science, the science of substance too would be a demonstrative science; but it does not seem that there is demonstration of the 'what it is'.

To which you might respond by saying that if there is not demonstration of the 'what it is', nothing prevents the same science from being both definitory and demonstrative, definitory of the 'what it is', demonstrative of essential accidents. For it belongs to the same science to know the 'what it is' of the universe, of the sun and moon, and to demonstrate their eternity and activities.

997a32-4 But if it belongs to another science, which will be the science that will know the accidents that relate to the substance? For it would be extremely difficult[128] to specify this.

[B] And indeed it would be most difficult to discover it. For if the science demonstrative of essential accidents goes as far as the first cause and the 'what it is', it will absolutely be definitory also. One should say however that such a science would be demonstrative in the proper sense, whereas there is already yet another science which demonstrates from hypotheses, which does not reach as far as the 'what it is', such as astronomy for the most part, wherefore it is not able to speak of the substance of the visible stars. Consequently, from all this is established what we have already said, that the science of substance is also demonstrative, but that not all demonstration ascends as far as definitory science, but possesses definition in such a way that it uses it as a principle on the basis of belief.[129]

Fourth Problem[130]

> **997a34-b3** Furthermore, one should say whether there are only perceptible substances, or also other ones besides these, and whether there be one sort or several genera of substances, as claimed by those who speak of Forms and intermediates, which they say are the objects of the mathematical sciences.

Aristotle placed this as the fourth in the series of problems.[131] But we have adequately answered the difficulty presented here when we dealt with these problems.[132] So [here] we merely indicate this much, [i] as regards the order of problems, that perhaps this should have been ordered as the third problem, and [then] [ii] we move on to discover the errors of the arguments that [seek to] destroy the position that there are several kinds of substances.

[i] For the *first* problem concerned the question if the wise man deals with all the causes of being, and the *second* problem concerned the question if he also deals with demonstrative principles. One might then have introduced plausibly, as the *third* problem, the question if there are many [kinds of] substances, the matter discussed now by Aristotle. Then, as the *fourth* problem, if there is one science of the many substances. Then, as the *fifth*, what here appears as the fourth, if there is one science which deals, not only with many substances, but also with their essential properties.

But let one judge these matters as one wishes; we then [ii] will go on to what follows.

> **997b3-5** Now the way in which we say that Forms are causes and substances in themselves has been set forth in the first discussions concerning them.

He refers us to what has been said in book 1. (This also shows that it is ridiculous for some to say that book 1 is inauthentic.)[133] But what he says there of Plato's views is nothing or very little, and that which he supposed [was Plato's view], he refuted there in a particularly empty and dialectical way.

> **997b5-12** These involve difficulty of many kinds, not least that it is absurd to say that there are certain natures besides those in the heavens, but that these are the same as perceptible substances, except that they are eternal whereas perceptibles are destructible. They say that there is a man itself, horse, health [itself] and nothing else, doing the same sort of thing as

> those who say there are gods, but gods just like men. For the latter do nothing other than produce[134] eternal men, nor do the former anything other, in positing the Forms, than produce eternal perceptibles.

But these men, I might say, hold that Forms differ in all respects from the things here below. For Forms are divine indivisible substances, self-reverting, the demiurgic causes of the things here below, eternally identical and the same, whereas perceptibles are subject to all types of change and some of them are mortal, they are effects and are found in the [realm of] coming-to-be that is involved in time. Forms are intelligible and indivisible and belong to themselves and <through>[135] this itself providing existence to all things in the universe, differing in their active, demiurgic, providential and epistemic power from all that which you, [Aristotle], describe as separate forms, whereas perceptibles suffer discord, are divisible, moved by others and the last and most sterile of beings. But there is a danger that Aristotle himself be caught in making excessive use of the deceptions of homonymy.[136] For it is not the case, since this man here is the same as man there, that they think these differ only in terms of eternity, and not also both in terms of all the differences we mention and in terms of yet many others which we have omitted.

One might be surprised to find that Alexander, also arguing in these matters with great insistence for what Aristotle says here, without noticing it, says the opposite of what Aristotle says.[137] For if man here is *synonymous* with man there, as Alexander's account has it, it would not even be by eternity that the one there would differ from the one here. For mortality is included in the definition (*logos*) of man here. And indeed in general in those [things] in which is present mortality and indestructibility, these are present [in them] substantially and not accidentally, as Aristotle himself has clearly demonstrated in this same treatise.[138] Thus, to follow Alexander's account, Aristotle did not correctly agree to this. But these men, I think, being so clever in logical arguments, do not differ in these matters, since they do not reach the divine views
24,1 of Plato, just as Thracian arrows do not reach the aetherial gods.[139]

But let us go on to what follows:

> **997b12-14** Furthermore, if someone posits the intermediates besides the Forms and perceptibles, this will involve many difficulties.

But let us see which difficulties, how many, and how they can be resolved.

997b14-18 For it is clear that there will be in like manner lines besides lines themselves and perceptible lines, and each of the other kinds. So that if astronomy is one science of these, there will be a heaven besides the perceptible heavens and a sun and moon and in like manner the rest of what is in the heavens.

And what is absurd in there being these [realities], intelligible, discursive and perceptible? It is necessary, is it not, that there be, in the demiurge, the cause of the heavens and of the sun? And that there be in souls a heaven and sun which are truer than those stars which are visible. And [finally] that there be these perceptibles receptive of them? What else do we think Plato is saying, when he shows the leader as doing astronomy 'above the heavens'[140] and then again 'many blissful visions and passages (*diexodous*) within the heavens enjoyed by the class of happy gods'?[141] Does he not think that there are stars above these heavens (or, speaking in more familiar terms, the intellectual lives of the stars) which he who is wise, according to Plato, contemplates in their unconfused unity, in their never-failing community and co-ordination with intelligible substances, and another intellectual heaven besides the visible one here, of which the sights and discursive[142] visions belong to the cosmic gods and demons? Consequently it has been very well said that not only will there be certain intermediate lines, lines besides these (i.e. besides the intelligibles and also besides perceptibles), but also that there are heavens and suns of many kinds. But all of these, to begin with, may be thought of at least as of three kinds.[143] For if, as the divine Plato says, the god, who made the whole cosmic living being intelligent and ensouled, placed 'intellect in soul, and soul in body',[144] then all those things that are seen in a perceptible and extended way in the visible heavens will be seen in immaterial and general reason-principles on the level of soul and in most part-less and intellectual forms on the level of Intellect.

But Aristotle, affecting to be amazed, writes as follows:

997b18-20 But how is one supposed to believe these things? For it is not plausible that it be unmoved, but absolutely impossible that it be moving.

For what reason is it not plausible that there be an unmoved cause of things which are moved? How can it not be necessary? Or, agreeing to this, do we[145] contend that the cause is not named in the same way as its products? And indeed we[146] have agreed on this as regards the products of art, maintaining that the enmattered house comes from an immaterial house. However one should also keep to the practice of the theologians, who reveal what proceeds from many heavens, from many suns, and from each of the many other [things in the heavens].[147]

But for what other reason is it absolutely impossible that this be moving, if it is not because we are accustomed to name as movement only bodily change? For who would not grant to beings, which truly are, demiurgic and intellectual motion?

997b20-2 And it is the same thing for those objects studied by optics and by mathematical harmonics. For these things also, it is impossible for there to be such besides perceptibles, for the same reasons.

Aristotle says that it is not plausible that they be either unmoved or moving. But then we said[148] that it is necessary that they be both unmoved, in regard to bodily movement, and moving, in regard to the movement of life. And now I would like most to ask him this: does he not agree that we have another eye, besides this visible eye? He grants that there is an eye,[149] but as not having the faculty to see the forms proper to it. Yet even granting this [latter point], [he would say] that what is seen by intellect has no existence. And will it not then be the case that intellect, which we[150] honour so much, will be less successful than perception, if indeed perception is ordered to things which are, whereas what intellect sees has no existence? But if this is absurd, and we have an intellectual eye, which will have a faculty to see, there will also be forms co-ordinate with this faculty, forms which are unmoved in relation to bodily movement, but are moving in relation to intellectual activity.

Come now! Is there no one who contemplates the harmonic principles which the god gave soul prior to the ordering of this visible world?[151] All the gods in the universe, do they not share in the Muses themselves and in the harmony that proceeds from them (gods full, as Homer says 'of the Muses singing alternately with sweet voice'),[152] whereas all beings share in them? Why therefore will there not be other objects of harmony? But if these sharp men[153] say that these things are Platonic and mythical, is it not still the case that the single co-ordination whereby you perfect the relation of your intellectual forms to the most primary good and the intelligible form within it,[154] is this [co-ordination] not the finest and most divine of harmonies, in dealing with which the philosopher would grasp objects of harmony more wonderful than those found in earthly music?[155]

997b23-4 For if there are intermediate perceptibles and perceptions, there will clearly be living beings between them and destructible beings.

Intermediates are not perceptibles even if perceptibles be homonymous with them, nor are perceptions co-ordinate with them [the intermediates], but rather discursive thinking, since these are immaterial reason-principles. Nor would one refuse to concede that between intelligible living beings and perceptible living beings there are other living beings in the universal reason-principles in soul or in natural forms. Thus what is supposed by Aristotle is <not>[156] true, nor is what is derived from these suppositions absurd.

> **997b25-6** One might be puzzled as to the kinds of beings these sciences should investigate.

Now this is a good question. For it really is a difficult point as to which sciences are more paradigmatic than the sciences of things here below. Nevertheless we will answer him that necessarily prior to the sciences that grasp natural forms, or reason-principles, or physical numbers,[157] are the sciences that deal with the immaterial paradigms of these.

> **997b26-32** For if geometry will differ from land-measurement merely in that the latter deals with what we perceive, whereas the former with what is not perceptible, it is clear also that there will be another science besides medicine and besides each of the other sciences, [a science] between medicine itself and this medicine [here]. Yet how could this be? For there would then be certain healthy objects besides those that are perceptible and the healthy in itself.

Land-measurement does not simply differ in this way from geometry, but also in that it aims at human use and sees to what is useful in mortal existence, undertakes manual tasks and does not provide a conception of true being, <whereas geometry leads away>[158] soul from perceptibles, purifies soul's eye and brings it around towards intellection,[159] separating it from the order which is more dense and shapes enmattered realities. Let it be said, nevertheless, that they differ in this way. What then? Does it not follow that there will also be another medicine, besides divine medicine and the medicine of mortal bodies? But I myself would be ashamed not to concede this, prior to all argument.[160] For let there be Paean or, if someone wants to say, Saviour Asclepius, who possesses divine medicine, by means of which, for gods, souls and bodies, he assigns the appropriate measure eternally to some, to others whenever they be receptive [of it]. Then let there be, as well, the doctor among us who cares for mortal bodies. Are then these [doctors], who differ so much from each other, bound by no common terms? And yet whatever may be found among the gods, or at the lowest level among us, must also be seen in the orders of the angels, demons and heroes and furthermore among souls that are released from mortal life. For what reason might one not also rank the philosophy which purifies [us] from affects between divine justice, which heals all vice related to the universe,[161] and human medicine, which merely takes care of the generated and destructible body?[162] Thus there would be nothing odd in there being matters of health known and practised by a medicine, [matters] intermediate between the measures of justice, in which health itself is found, and those things recognised by most people. But we, searching, not for the paradigms of things here below, but for the things visible at the lowest levels in the changings of things, and not finding [them],[163] we are puzzled, no wonder, as to whether or not there might be such among the primary and intermediate genera of being.

Now as to the question whether we need to posit paradigmatic causes for all the arts, or if the principles of all the arts derive from above, whereas the whole set of arts which contribute usefully to mortal life has proceeded from the generative reason-principle of soul: these questions were examined both by the most ancient philosophers, I mean by the Pythagoreans, Parmenides and Plato, and by those who later explicated their knowledge, Plotinus, Iamblichus and Porphyry.[164] However, that we may conceive of corresponding sciences as also among the gods is also made clear by the theologians, who do not hesitate to call the Cyclops and Hephaistos coppersmiths and builders, and to assign the art of weaving to Athena and Kore. We say that these corresponding sciences are also the paradigms of things here below, but not without mediation, unless [in the sense that] the arts here operate in accordance with the divine sciences, tending towards them and imitating them.

> **997b32-3** This also is not true, that land-measurement deals with perceptible magnitudes that are destructible, for it would have been destroyed with their destruction.

It is especially Aristotle himself who supposed that they say that land-measurement deals with destructibles, and we do not find that this is in their [philosophy]. Furthermore, why is it necessary that the arts dealing with destructibles be destroyed with them? It is as if one said that the definitions of destructibles are destroyed with what they define. But I think this: the form surviving, the definition necessarily survives, even if particular things are mortal, whether or not one expresses [the definition]. The same indeed obtains also in relation to the arts: for they survive, even if their concern (*pronoêtikas*) is destructible things, since they are controlled by universal reason-principles. And even if these are projected by souls which find themselves in coming-to-be, arts of this sort also must remain. For what is destructible differs in any case from what cares (*pronoein*) for the destructible or acts in relation to the destructible: that which is in the destructible necessarily is destroyed with the substrate,

whereas that which acts in relation to the destructible, when this is 27,1
destroyed, is not necessarily no longer existent or idle. For it will act again in relation to something of the same kind through using a universal reason-principle.

> **997b34-998a6** But indeed the astronomy of this heaven[165] would not be about perceptible magnitudes. For neither are perceptible lines such as the geometer says (for nothing perceptible is straight or curved in this way, for the circle does not touch the straight-edge at a point, but in the way Protagoras indicates in refuting the geometers), nor are the motions and orbits of the heavens such as those discussed by astronomy, nor do points have the same nature as stars.

And yet if the astronomer does not consider perceptible magnitudes, but deals with stars and the heavens and the motions of these, why are there not other stars, another heaven and other circular orbits, with which the astronomer deals? Thus the objection is not against those who introduce several kinds of substances, rather it agrees with them. However the truth is as follows: the astronomer considers this heaven, but since he has within himself universal principles which are images of the forms which produce the heavens, he combines with the visible things the immaterial and more comprehensive reason-principles whence he demonstrates the essential properties of the heavens. See how the heaven, although it is replete with life and with divine demiurgic power, nevertheless, on account of its mass and extension, receives things as divided in parts (*merikôs*). For there is, in us, one reason-principle dividing all the major circles in the sphere,[166] which is an image of the reason-principle which with Zeus brings the universe into existence. But the heavens receive this reason-principle, here as in the zodiac and equator, there as in the meridian and the horizon. For bodily nature does not allow that all the power of the reason-principle be received simultaneously and in one [thing]. I would also say the same as regards the other points. And it is no wonder that it [the heaven] receives [the reason-principle], not only separately, but also not simultaneously and not as accurately as it is given by that which, coming directly from the Father,[167] makes and [as accurately] as it can know [it].

But against Protagoras and against all those who devalue geometry and those who say that there exists only perceptible things, this is to be said: are the curved and straight [things] in the heavens and in general in the perceptible [world] more precise than that which the geometer considers?[168] For if what is in the perceptible is more precise, then it is strange that those things which do not fit definitions are finer and more precise than those which do, and that those things which are determined not to be curved by means of a straight-edge[169] should contest pride of place with those which do measure up to it. But if those objects which geometry <considers>[170] are superior in immateriality, purity, precision and truth, are not the following two points shown?: that geometry concerns worthy and valuable kinds of things (*genê*); and that superior and more divine paradigms of what is in[171] the heavens pre-exist in the demiurgic intellect and in soul as a whole, which our constituted nature (*sustasis*), having images of them, knows, encompassing both what is in the universe by means of more universal reason-principles, and what is in [the soul's] most primary causes, ascending thus through the images in it to the archetypal substance.

28,1 But it is also absurd to refute geometry by taking it that points are not equal to stars. For neither do astronomers say that stars are points, nor does geometry seek to take as points such things which belong entirely to the perceptible. But if one were in general to refute in this way the geometers, on the grounds that that of which they speak is not visible in perceptible things, one should examine the poles or the centre of the universe.

But let these things go, and let us examine Aristotle's correction of the last opinion of those who introduce several [kinds of] substances.

> **998a7-19** But there are some who say that there are these intermediates which are said to be between Forms and perceptibles, but that they are not separate from perceptibles but in them. All the impossible implications of this would require a longer discussion, but the following considerations will suffice. For it is not plausible that this applies only to the intermediates, but it is clear that Forms could also be in perceptibles (for the same account concerns both cases). Then again it would be necessary that two solids be in the same place, and would not be unmoved, being in perceptibles which move. And in general for what reason would one posit these things as existing, yet as being in perceptibles? For there will be the same absurd implications as those said before, for there will be a heaven besides the heaven, yet not separate, but in the same place, which is most impossible.

As to those who say there is some intermediate thing between intelligible Forms and perceptibles, and that this intermediate nature is in the lowest, enmattered forms, if they say that the whole of the intermediate is in perceptible particulars, they admittedly err. But if they say, given that one can in several ways conceive of universal man, say, and the sun (one being in the soul of the universe, another in the cosmic nature, another in our souls), that there is a common man also co-ordinate with Socrates, and that the sun itself, even if it is one, has therefore the quality of the form [of a sun] as a whole, they do not seem to me to go wrong, since this relates to the truth of the matter. And it is often to be heard from those who are inspired by Aristotle's philosophy that the common is to be found *in* particulars. Even the Stoics put common qualities in[172] the peculiarly qualified.[173] What else is what is defined but what is common and co-ordinate with particulars? Thus if one is to define the sun, the heavens, or the moon, one might define each of these according to that which would be present in all suns, which, even if there were thousands of them, would be similar in form to each other.

This is the way things are. But let us see if those who say that each perceptible thing has something universal co-ordinate with it have anything to be ashamed about in the discussion. Well then, he asks them why they do not also put the Forms in perceptibles. But I think a solution to this lies at hand for them. For they will say that the Forms are immaterial, entirely separate, intellectual, demiurgic. Belonging only to themselves, they generate and bring to existence all things, whereas these[174] which are universal in this sense [as common] tend towards matter, are not able to remain in themselves, becoming receptive of the 'peculiarly qualified' and have existence together with them.

As to Aristotle's objection as to how two solids can be in the same place, they will deny that the universal reason-principle is a solid: it contributes to the one substance of a solid body, but is not itself another solid besides the body. So Socrates has within himself both the co-ordinate [universal] man, and the peculiarly qualified. And there are not two animals in him or two men, or two solids, but all these come together to constitute one nature.

But how, Aristotle says, do the unmoved yet remain such in being in what is moved? Is it not the case that they are unmoved as long as the thing exists, since neither do the peculiarly qualified move if the thing exists, but are not simply unmoved, if they are in a generated thing? For with Socrates are destroyed the man in him and the peculiarly qualified.

In addition to all this, Aristotle questions why it is necessary both 29,1
to introduce these things and, in introducing them, to put them in perceptibles. To this one should reply that since perceptibles are imitations, they must have images in them of intelligibles and themselves have the more comprehensive and more common as prior to the more specific and partial. For in the way that, in immaterial Forms, the more powerful and comprehensive is always prior to the more particular and subordinate, thus also in material things is the common quality (*to koinôs poion*) presupposed by the peculiarly qualified, it being a product and image of the more comprehensive reason-principle, but also a formed place (*khôra*) receptive (*hupo-*

dokhê) of the peculiarly qualified. There is therefore nothing surprising in there being in the heavens a common substance of the heavens, according to which this heaven is defined, and which the heavens, were there thousands of them, would have, and also that which is the shape and distinctive property of this heaven.

Chapter 3
Seventh Problem

998a20-b3 Thus there is much difficulty concerning these matters and what should be posited so as to reach the truth. And, as regards principles, whether genera are to be taken as elements and principles, or rather the primary components of each thing, such as the elements of speech. And the principles seem to be those things of which all sound[175] is primarily constituted, not what is common, speech. And we say the elements of geometrical propositions are those things of which the demonstrations are found in the demonstration of other things, of all, or of most. Then again, regarding bodies, those who say there are several elements or one from which bodies are composed, they say these are principles. For example, Empedocles says that fire and water and what comes after these are the elements out of which beings are composed. But he does not say these elements are the genera of beings. In relation to this, if one will look also at the nature of the other things, for example of a bed, [when one knows] the parts out of which it is[176] and how they are arranged, then one will know its nature.

The sixth in the presentation of the problems raised by Aristotle concerned him who treats in a scientific way of the same, the similar, the opposite, the different and the dissimilar, just as the dialectician deals with these on the level of opinion.[177] But Aristotle here passed over this problem, perhaps because he intended in the next book, book 4, to show clearly that these matters are the appropriate concern of him who does first philosophy (and indeed the topic did not require a long exploration, but rather a statement and brief judgement), or because he will also provide us in book 10 with a complete teaching on these points. Thus, for reasons such as these, passing over this, what he at first [ch. 1] marked as the seventh problem, he now takes up as the sixth and enquires whether it is the components of things which are principles, or genera. But we have already answered[178] as to which principles are components, and which are genera (genera taken, not in the sense of those which are conceptions developed *a posteriori*, but those which are primary active principles, established in the causes of all).

However we do not find anything obscure in Aristotle's presentation of the problem, nor in his argument [A] that components are principles. Hence let us move to what follows, where he says [B] that genera are principles to a greater degree.

998b4-6 But if we know each thing through definitions, and genera are the principles of definitions, it is necessary that genera are the principles of what is defined.

It is true in a way that we know each thing through definitions. But how are genera the principles of definitions, if it is not as components of definitions? Therefore through this also we demonstrate that the components are principles to a greater degree than anything else. For if genera are principles for the reason that they are components of definitions, even if we wish to argue, as it seems, that a principle is something other than components, we come back again to components. Unless indeed genera are principles of definitions in this way: that definitions are species, and genera are predicated of species. But if we accept, among genera, those which are *a posteriori* and generated additionally,[179] how could we say that these are the principles of species or of definitions? However, if we would think of the genera which preside in the universe over the organisation of beings according to species, perhaps in this way we might come to principles in a truer sense.

998b6-8 And[180] if it is to be taken that the science of beings is the science of the species according to which beings are said, genera are the principles of species.

It has been said which genera are in reality the causes and principles of species.

998b9-11 Some also of those who say that the elements of beings are the One, or Being, or the Large and Small, seem to use these as genera.

Those who say that the One or Being or the Indefinite Dyad are principles of beings have also shown them to be perfective principles, intending the good to come to all beings from the One, being to come from Being, and, from the Indefinite Dyad, powers and activities and movements and proceedings and growth of all sorts, there being nothing among beings which does not share in the One and the Indefinite Dyad and Being and is not generated from them. If therefore, for this reason, one wishes them to call these the genera of all things, we will not quibble over terminology.

But having said (998a20-b3) for what reasons components are principles to a greater degree than genera, using Empedocles as a witness, Aristotle will say why genera are principles to a greater degree than components. Advancing the Pythagoreans and Plato as advocates of this view, it now remains for him to say that one or other of these positions must prevail, for both cannot say the truth. But we have said before[181] that both can speak true, if at different times different principles are assumed.

> **998b11-14** But indeed it is not possible to speak of principles in both ways. For the definition of substance is one, but the definition by means of genera is[182] different from that which says out of which components it is.

Aristotle has shown in the *Posterior Analytics*[183] that for each of the species the most proper definition is one, and that this definition is wholly through genera, so that it is in this way especially, following his judgement, that genera might be said to be principles. For if a definition comes from genera (for differentiae are, like species, related to genus),[184] and principles are that out of which comes the definition, then genera would be principles. Unless those who represent the opposite position show that genera and differentiae, out of which definitions are made, are elements of the species, remaining and present in the species as being co-ordinate with it, not as transcending the species, nor as their efficient or final causes: in this way, again, the components would be principles. So much for this war of words. But the true position has been spoken,[185] on the basis of which one should arbitrate between the conflicting arguments.

Eighth Problem

> **998b14-16** In addition, if genera are principles to the highest degree, whether one must think that it is the primary or the ultimate genera predicated of individuals that are principles.

This is the eighth in the order of problems,[186] but since the sixth has been left aside, it is the seventh now to be taken up for practice
31,1 (*gumnazetai*). But we have already spoken about this[187] and the truth of the matter may be derived from there. Yet let us now see what force the opposing arguments might have.

> **998b17-21** For if universals are always[188] principles to a greater degree, it is clear that the highest of the genera [are principles], for these are said of all things. Therefore there will be as many principles of beings as there are primary genera, such that being and unity will be principles and substances. For these, most of all, are said of all beings.

It would not be opportune for either position to speak of the non-existence (*anuparxia*) of universals, for both positions take it that universals exist, one group being keen on conferring on what is most general the dignity of being principles, the other group thinking rather that what is most specific is more suitable for the position of being a principle.

Well, the argument that advocates what is most general, given that the notion (*ennoia*) of principle includes what is most eternal, what is immaterial, what is most encompassing of beings and transcends and is separate from the lowest of things, judges that what is more universal is more a principle and that what is most general is of this kind. Hence it [what is most general] is said of all things as if encompassing all in one nature and capable of being present to all things while being separate. Now if this position were to prevail, there would be as many principles as there are highest genera, whether there be ten, or six, or five, or two [of these]. But it would seem that what would be best is the position that there are unity and being, for there is nothing among beings of which neither of these two will be said. This then is what these people tell us.[189]

But [A] the argument opposed to them will try to show, first, that these [two] (I mean being and unity) are not genera; then, that they are not principles; and then, in this way, that there are not other things that are by nature the highest genera. For being and unity are not genera to the highest degree, if indeed they are not genera. He therefore argues that what truly are genera to the highest degree are not principles, beginning from this point:

998b22 But it is not possible for either unity or being to be one genus of beings.

I have already said how the ancients said that unity and being are principles which most encompass beings. Now the argument shows that these [unity and being] do not admit of the distinctive property of genera generated additionally.[190] It begins thus: unity and being are predicated of differentiae; species and genera are not predicated of differentiae; unity and being are neither genus nor species. He explains these points in this way.

998b23-4 For it is necessary that the differentiae of each genus exist and each be one.

That is, it is necessary that unity and being be predicated of every differentia.

998b24-6 For[191] it is not possible to predicate either the species (of a genus) of its proper differentiae, or a genus without its species.

It is as if he were saying: it is not possible to predicate man of rationality as a whole (for this differentia is proper to him and formative; but how might what is lesser be predicated of what is greater in extension?), or animal of rationality, as if rationality were somehow to lie outside human nature. Thus, neither would man, which is the species of a genus, ever be said of his differentia, nor would animal, in not being predicated of man, ever be predicated in particular of rationality. However, unity and being are predicated of every differentia. Therefore unity and being are neither a species nor a genus of beings.

998b26-7[192] So that if unity or being were to be a genus, no differentia would be either being or one.

But if someone were to insist on saying that unity and being are in every case genera, it follows for him that these are predicated of no differentia, which is absurd and completely overturns the supposition of those who advocate these. For then unity and being would no longer be what is genus to the highest degree, nor will they be predicated of all things.

These are the [dialectical] criticisms that Aristotle here develops. However, one should know that the whole argument about differentiae requires correcting as regards these men.[193] For they do not teach us whence differentiae enter into the forming of things, since they deny that differentiae are contained in genera, nor do they show in a way worthy of note how it is still possible to predicate a genus of a species as a whole, if it does not include differentiae, nor, having examined the differentiae of genera – some in relation to the unco-ordinate, some according to the co-ordinate, some according to what is prior among the species, some according to what is posterior (*neôteron*) – have they distinguished adequately what genera have differentiae also in act, what [genera] underlie species as if having the function of matter, which of them in precontaining their primary productive cause produce their complete constituted nature, which of them as imitating the true causes of species are of a nature such as to be merely predicated of them.

But the most wrong-headed of opinions, it seems to me, is that which Alexander presents in the discussion of these points,[194] as if having thought up a cure, it seems, for the doctrine at issue. He says that as regards trees, for example, the differentiae are not in the plant, nor are the differentiae of a plant in the animated body, nor are the differentiae of the animated body in the body, yet all the differentiae of substances are in substance: animal does not have the differentiae of man, nor does the animated body have the differentiae of animal, but substance has all the differentiae of animals and of animated bodies. But if animal does not have the differentiae of the species subordinate to it for the reason that it will not be in its activity at the same time irrational and rational, mortal and immortal, why will the same not happen again as regards substance? How, in dividing substance, will you not divide it also into differentiae, but only into species? Why will animal not have the same relation to man as substance has to animal? So that if substance is predicated of the differentiae which are formative of an animal, so animal will be predicated of those that are formative of man. Again, if substance [is predicated] of the differentiae, the differentiae are its species, so that they also will have differentiae whereby they are separated from the other substances. For every species coming under a genus is constituted by the genus and differentiae. And if the differentiae of differentiae come under substance, we will go to infinity, for the differentiae will again be species of substances and will have other differentiae. But if the differentiae are not substances, whence will they come to substances? For it will not be from another category or by chance. But how can it be that, if Aristotle says here (998b24-6) that the genus is not predicated of the differentiae without the species, we say that differentiae that are not in the species of substance are of themselves in substance?

Now, let these points as regards his doctrine be taken as settled, for they have required on our part an explanation going beyond what is proportionate. But we add again only this much, that the account of differentiae must be articulated, since these people have brought it into great confusion. Let us go on to what comes next.

> **998b27-8** But indeed if they are not genera, nor will they be principles, if, that is, genera are principles. 33,1

If genera are the only principles, it follows, according to the second [syllogistic] figure, that every principle is a genus; unity and being are not genera; unity and being are not principles. If however not every principle is a genus, but some principles are genera, the conclusion will be particular.

> **998b28-30** Then again the intermediates taken with the differentiae will be genera, going as far as the indivisible [species]. Now some seem to be this, others not.

If, he says, all that which is predicated, such as unity and being, is a genus and hence a principle, then all those which come between these and indivisible species, what we also call subaltern (*hupallêla*), will rightly be spoken of as genera and principles. Now they do not hold that all intermediates should be called genera. For they speak of those which have names as genera ('bird', 'animal', 'body'), but no longer call genera those without names (such as 'winged animal' or 'animal with feet'), since they are signified by means of two names. Again, whenever they divide genera by contradiction, they make the species from affirmation and from the genus, [the species] which is the genus of what comes after it, but they do not do so in using negation. For if, for example, they divided animal into what has no feet and what has feet, they undoubtedly produced 'animal with feet', but have <not>[195] looked into the matter if 'animal without feet' is incapable of being a formative principle. Is it not therefore the case that they set things up which conflict with themselves?

It is possible to examine in a general way (*koinôs*) what Aristotle say here, whether he is in earnest or using verbal (*logikôs*) refutations. But it is clear to those who have not forgotten this that, in the first [book] of *On the Parts of Animals* (1.3), he puts the same objections to Plato's method of division.

> **998b30-999a1** Then again differentiae will be principles to a greater degree than genera. But if differentiae are principles, then there will be so to speak infinite principles, especially if someone were to posit the first genus as[196] principle.

[This is against][197] the position that the most general are principles. For if genera are principles, then much more so are differentiae. For genera have the function of matter, differentiae the function of form, but form is more a principle than matter. Therefore, if genera are principles, then much more so are differentiae. But if differentiae are also principles, then there would be almost infinite principles. For if one were to think of the species subordinate to substance and [of the fact] that differentiae are more numerous than species, then he will know that they will be entirely unknowable for us on account of their multiplicity. For who will be able to go through the variations of all plants, stones, trees and animals? For they are probably, by their proper nature, near to infinity. But principles must be as few as possible in number, in power being generative and encompassing all things.

This then is the nonsense that these people speak. But if one were to say that all differentiae are encompassed in a unified mode in the most general of causes, he will escape the numerical multitude of principles and will say[198] that nothing in the universe is generated without cause.

34,1 **999a1-6** But if however unity is more of a principle, and unity is the indivisible, and all indivisibility is according to quantity

> or species, but indivisibility according to species is prior, and genera are divisible into species, what is last predicated must be what is most one. For man is not the genus of particular men.

The argument wants to say that the indivisible species, as being indivisible and more one, is a principle, rather than the genus. For it argues that the genus is divisible and not more one than many. For since a division can be made in two ways, and even if division according to species is prior (for it also divides substances that are entirely incorporeal as well as the ten categories, whereas the other [mode of division] concerns quantities only), according to each [mode of division] what divides will be less of a unity than that which is divided, and most especially those which divide according to species, for this produces many wholes. Therefore genera divided into species contain within themselves a plurality, whereas the most specific form is not predicated <like a genus>[199] of particulars (so it is not divided into them), but is as it were their indivisible species.

He constructs <the argument>[200] thus: what is most specific is more one than the genus; what is more one is more a principle; the most specific is more a principle than the genus. But the argument goes wrong in relation to the minor premise.[201] For it is not possible for what is near things which are infinite to be more one than what is situated at a greater distance from them. And in general genus is not thus divided into species so as to be used up in their coming to be, but true genus staying what it is, species come to be. Therefore, in power genus is more comprehensive than they are, but is more one in number: but this, we say, is appropriate for principles.

> **999a6-13** Then, where there is the prior and posterior, what is predicated of them cannot be something besides them. Thus, if the dyad is the first of numbers, there will not be a number besides the species of numbers, nor in the same way will there be a figure besides the species of figures. But if it is not the case for these, then much less so is it the case that the genera of others will exist besides the species, for it is especially of these that there seem to be genera. But there is no prior and posterior in indivisible [species].

This critique appears to destroy even the existence of genera, in such a way that only indivisible species will exist, and which will be accepted for the function of principle. However, it does not achieve this, and were it capable of this, species would also be immediately destroyed with genera. For they exist and are said in relation to each other. Where then does the argument work and where does it go wrong?

Now it is agreed, even by us, that there is the prior and posterior,

the better and worse, when it comes to species of the same genus. But now, in order to avoid adding confusion to the topic by comparing celestial, divine, immortal and intellectual animals with those which are terrestrial, mortal, destructible and irrational, I believe the discussion (*logos*) should deal with all the latter. If indeed differentiae, according to which we distinguish the powers of species in relation to the nature of genera, are not of equal value and in many cases have among themselves an order in relation to form and privation, it is most necessary that species also differ from each other in relation to the better and worse, to the prior and posterior.[202] Indeed, for the Pythagoreans, the odd is of more value than the even, and the circle than rectilinear figures, and again the body having circular motion than that having linear motion, and all the rest are in general like this. But it is not for this reason that these species are
35,1 not ordered under one genus. For (i) either the differentiae are not in the genus, but are added to the species from outside, and the species have the same and the similar from the genus, the prior and posterior, the better and worse from the differentiae; or (ii) the genus will also contain all the differentiae of what it has in itself, and it will produce many species, some being nearer to it, some having an intermediate rank among those which proceeded from it, some appearing at the end of the structures that are generated from it. But in all of these the genus places an image, mark and reflection of itself, through which they can return back to their single and most general principle. For they are not prevented from also being univocal with each other, with respect to this single reflection, even if, in relation to other causes, they have in respect to themselves much difference and otherness.

Thus in general the argument is reversed[203] as regards attempts aimed at doing away with the existence of genera. For some [genera] are shown, through their superior power and as most generative cause, to produce the primary, intermediate and last species. And some, those species said of individuals, have come to be almost like individuals themselves in being ranked very near matter, which is by nature infertile, constricts whatever approaches it[204] and does not allow it to extend to a multitude of formal powers. For this reason the indivisible species cannot be predicated in respect to the prior and posterior, of the better and worse, on account of its very limited nature and the limit of what is proper to it.

999a16-17 But then again it is not easy to say how these are to be understood as principles.

[B] Aristotle wants, then again, to argue on behalf of the opposite position and to show that it is not indivisible species, but the most general that are principles. This is how he argues: the principle of beings and the true cause should be separate from beings. Why then is the indivisible species separate from particulars? Is it not because it is universal and said of all? But in that case genera, since they are more universal and predicated of all, are separate to a greater degree from sensibles. Consequently, genera are principles to a greater degree.

And indeed, [we say,] both the axiom that causes should be separate is entirely true, and the entire development of the argument is sound: it is demonstrative, but demonstrates from a hypothesis. For <if>[205] indeed indivisible species are principles, even more so are genera principles. For in general it will be necessary to acknowledge as applying to a greater degree to genera whatever axiom is applied to species. Thus someone who says that species are insubstantial, as generated *a posteriori* (*husterogenê*), will say the same[206] with greater force as regards genera. For this reason also both species and genera are [said] in the *Categories*[207] [to be] insubstantial, but the genus more so than the species.

Chapter 4
Tenth Problem[208]

> **999a24-32** These matters include a difficulty, the most difficult and most necessary to examine, to which the argument has now brought us. For if there is nothing besides particulars, but particulars are infinite, how can there be science of the infinite? For we know all things in that there is some one identical thing, and in that there is something universal. But if this is necessary and there must be something besides particulars, there must be genera besides particulars, either the last or the first. But we have just explored the impossibility of this. 36,1

The order of this topic too has been changed. For what was tenth in the [presentation of the] problems[209] is taken up as the eighth in the argumentative exercise (*gumnasias*). Aristotle enquires whether or not what is said universally exists besides perceptibles. If not, he says, then there is no science of beings. For it will not be possible that we learn of these things which come to be an infinite number of times in an infinite time. We think now we can grasp them through universals, but if these are non-existent, then definition, demonstration, division <and indeed>[210] all philosophy, haunted by vain dreams, will vanish. But if, he says, there are universals, then there will be genera and species which, we have just finished saying, do not exist. And the reason, it seems, why he connects this question with the seventh problem is that he wishes to remind us, through its proximity, of what he said there.

Now let us say that it has not been shown either, in the previous chapter, that genera and species do not exist. For how could what is false ever be demonstrated? If, however, universals do not exist, science will disappear from things. And indeed you [Aristotle] do well to say this. But add that what is infinite also could not come to be, if there were not one cause which is infinite in power, unless it were to be spontaneously. But how could things [existing] by nature come to be spontaneously? Consequently, if it be thus, and infinite men and horses and many univocal things come to be in an infinite time, for each of them there is a henad infinite in power, according to which these come to be in a determinate way, to infinity, in the universe, and are known <by those who know>[211] the cause of that which makes infinite things. Furthermore, all animals that are generated change from potentiality to actuality. If so, there must be in the universe an animal in actuality which brings animals in potentiality to actuality. But this is neither the sun, nor the moon, nor any of the other animals already determined (for the divine animals[212] contribute also to the coming to be and perfecting of what is generated),[213] but is that *by* which in a primary sense (*kuriôs*) and *for* which things here below come to be. This will be an animal in actuality only, and nothing else, that is the cause of animals ranked as individuals. It is in this way that, for me, you should consider man and each of the other [animals] and you will see that the true genera[214] are naturally generative and that true species are formative of what is infinite.

> **999a32-b1** Again if it is especially the case that there is something besides the composite (whenever something is predicated of matter), whether, if it is a form,[215] it must be besides all things, or besides some, but not besides others, or besides none.

By 'composite' here, Aristotle means particulars, such as what is formed and has a form in a substrate, which he says is predicated of matter. For it is his habit to call what is in a substrate as [predicated] of a substrate. His division is complete: for universals exist either besides all perceptibles, or besides some, or besides none. But since he starts the division from a hypothesis (if there are universals, he says), the third option destroys the hypothesis, just as Alexander before us has indicated.[216] Perhaps then the hypothesis is to be understood in this way: if we should consider in general if there be universals, we will say that there are such for all things, for some, or for none. Then [Aristotle] shows that to do away with them entirely is extremely destructive of everything.

> **999b1-4** Now if there is nothing besides particulars, there will be nothing which is intelligible, but all will be perceptibles and

there will be science of nothing, unless one says that perception is science.

On this hypothesis the most divine of beings would quickly disappear from existence and it seems that the god[217] would have given us intellect in vain, nor would there be science among us or among the gods, unless someone, like Theaetetus or Protagoras, wants to call perception science.[218] But this would somehow be a mockery and not something serious.

999b4-5 Furthermore there would be nothing eternal or unmoved (for all perceptibles are destroyed and are in movement).

He who destroys intelligibles and those things known by discursive reason[219] not only destroys intellect and the intelligible and science, but also all that is eternal. But why? Is not the heavens eternal? Is it not eternal rather because there is something intelligible and unmoved that is the cause of its eternal movement? If there were not this cause, what would be left but that 'the entire heavens and all coming-to-be in collapsing will come to a halt and will never again have that which, in moving it, will bring it into being' (*Phaedrus* 245D7-E2).[220] Of course Aristotle has also argued this point elsewhere, showing that all body is limited.[221] But now he will show us clearly that if there is not something that is eternal of its own nature, not even generation will come to be.

999b5-6 But if there is nothing eternal, nor is coming-to-be possible.

Let us learn from Aristotle's excellent demonstration why this is and on account of what necessity.

999b6-8 For that which is generated and that out of which it is generated must be something, and of these the last must be ungenerated.

If there is coming-to-be, there will at some time be what is generated (for change in the form of coming-to-be is not unterminable) and there will also be a material cause, which, when it is thought of as proximate, may be generated (for indeed the proximate matter of our body is generated), but which, if analysed to the end and brought back to prime matter, is of necessity ungenerated. For what reason? Because it is not possible to analyse the material cause to infinity, nor can anything be generated from what is not. And this is what Aristotle shows with [the words]:

999b8 If it is to stop and it cannot come from what is not.

Aristotle has shown in book 2 (ch. 2) that the analysis stops and does not go to infinity, taking it that there would not be a cause at all if there were no first cause. But if there is a first cause, then the analysis of causes, in going up, stops and does not go to infinity. It is a common axiom held by those who study nature that nothing in nature comes from what is not.

999b8-12 Furthermore, if there is coming-to-be and movement, it is necessary that there be a limit (for no movement is unlimited, but for all [movement] there is an end, and it is not possible that that which cannot be comes to be, but what came to be must be when it first came to be).

It is necessary that there be a limit for all coming-to-be. For natural movement is not without finality (*askopos*), nor is coming-to-be produced in nature as unterminable. But if all coming-to-be has a limit, and all that which has a limit also has a principle (for this must be implicitly understood as well), [then] all coming-to-be and movement has a principle and limit. But is this also the case for heavenly [motion]? It is clear that it is so. Since this movement occurs an infinite number of times, joining the end to the beginning, the heavens remain in eternal motion on account of the power which moves them. As to all that which is generated having a limit, Aristotle argues for this in this way. What may come to be comes to be; so all coming-to-be belongs to what has a limit; but if it belongs to what has a limit, then it belongs to what has a principle; so all coming-to-be and movement belongs to what has a principle and a limit. There is no limit or principle for that which is eternal; so no movement or coming-to-be belongs to the eternal. So if there is something which is eternal, it must be ungenerated and unmoved. Now this must be intelligible or the object of discursive thought,[222] for these are ungenerated and unmoved. So intelligibles and objects of discursive thought exist. But this is what [Aristotle] set out [to show] from the beginning. Let us then summarise all of this: if there is coming-to-be (but there must be), there is what is eternal; if there is what is eternal, there is something ungenerated and unmoved; if so, then it is intelligible; if so, not only perceptibles exist.

Ninth Problem

999b12-16 Furthermore, if there is matter because it is ungenerated, much more so is it plausible that substance exists, which matter at some moment becomes. For if there will be neither the former nor the latter, there will be nothing at all.

> But if this is impossible, there must be something besides the composite, the shape and the form.

This is the ninth in [the presentation of] the problems, as it is ninth here. The reason for the change in the order [of the problems] emerges clearly here. For since Aristotle had shown, in the tenth of the problems there,[223] and in the eighth here,[224] that matter is ungenerated, he joined to it immediately the argument concerning separate form. For if there is always matter on account of it being conceived as prior to all coming-to-be, then it is much more plausible to conceive as prior an ungenerated form which brings to actuality, to perfection and a formal order a matter which is in potentiality, is imperfect and has its being in the privation of all things.

[A] The argument therefore goes like this. If matter is ungenerated, much more so is form separate and eternal. For matter will not form itself, as was thought by some of the ancient philosophers of nature. But if someone denies that matter is ungenerated, so as not to be forced to posit form as ungenerated, he will learn that if neither is ungenerated, then there will be nothing. For it has been shown that there must be something eternal, if there is to be coming-to-be. But if it is not possible for there to be nothing (for what would be more impossible than this?), there must be something eternal. Now if this be form, then I have what I seek. But if it be matter, then I will say in turn that much more plausible also than there being ungenerated matter is that there be eternal form. So that since it is impossible that both not be, and, that if there be ungenerated matter then form must be, then it is necessary that both be. He has shown clearly that the argument concerns separate form since he adduces: 'there must be something besides the composite, the shape and the form'.

> **999b17-20** But if again someone will posit this, there is a difficulty as to for what things he will posit form, and for what things not. For it is clear that it cannot be for all things, for we would not posit a house besides particular houses.

[B] The difficulties which the followers of Parmenides[225] habitually raised are investigated here by Aristotle: of what things is there a Form (*idea*) and of what not. Now an accurate treatment of these matters is given in the divine Iamblichus' comments on the *Parmenides*.[226] There is also [a discussion] in Plotinus, in his study *On Intellect and the Forms and Being*.[227] But so as to say [what we think] 39,1
here in a few words:[228] one should not posit Forms of artefacts (Aristotle here rightly denies that there is a house besides the many houses), nor of things against nature, nor of what is evil or ugly, nor of particular generated things. We say that there are eternal paradigms, which are intellectual, generative and providential, presiding

over the whole forms of coming-to-be, the forms in the whole universe and the immaterial reason-principles of the soul.

> **999b20-3** Furthermore, will there be one substance for all, for example for all men? But this is absurd, for all will be one of which the substance is one. But will they be many and different? But this too is irrational.

One might also put this difficulty to those who introduce the Forms: if there is one Form for all men, what they describe as 'Man itself'. But he says that this is absurd. For then all men here below would also have to be one, since related to one: 'for all are one of which the substance is one'. We will ask: what substance? Substance in the sense of inseparable form,[229] or substance which is separate? <If in the sense of inseparable form,>[230] we say that the criticism is correct, but that the argument here is not about inseparable form. If in the sense of separate Form, then we will say that there is nothing absurd in there being generated from one cause many, indeed infinite things which, however, are similar to each other in form, if coming to be in relation to it. If then this is the correct text [in Aristotle], nothing absurd follows.

But if one were to choose to write: 'But this is absurd, for the substance of all is not one'[231], then we will say again the same things: that nothing prevents their separate substance and cause from being one, but that the inseparable [substance] is one in form, but many and infinite in number. It is in some sense irrational and in some sense necessary to posit many, separate and different causes of the same thing. For it suffices to posit one paradigm of men in the demiurgic Intellect, and absurd to posit several, since one suffices to generate all men. Since what will now distinguish between men, if they are in the same?[232] But there must also be a cause of men here below in the world-soul and in the nature of the whole, and in general forms for each rank of beings, as is shown elsewhere. But this whole enquiry arose from the problem as to whether there are Forms also of particulars, or none for any of the individual generated things.

> **999b23-4** Also, how does the matter at the same time come to be each of these, and the composite is both of these?

The way in which Forms mix[233] is indeed a source of difficulty. The divine Plato himself also says this in the *Timaeus*[234] and shows it in the *Parmenides*.[235] For does each of the things here below share in part of the Form? And how will the Form be divided, or these things be like each other in form, if they receive existence in relation to different parts of the Form? But is the Form present to each thing as a whole? And how does the one become many, but is shown <not>[236]

to be separated from itself? And in general all the difficulties which the great Parmenides has presented[237] in exercising (*gumnazein*) 40,1
Socrates in relation to the study of the Forms. Almost the same difficulties Aristotle raises in this chapter, difficulties which the friends of the Forms[238] were also in the habit of investigating:[239] if there are Forms, of what things there are Forms, how matter shares in them, by mixing, composition, alteration, intertwining, what makes matter and form one. For these are the habitual difficulties. But here in this work Aristotle himself says[240] that what is potentially perfected, going towards that which it is in its nature [to realise], is not in need of something to bind it to its perfection and actuality.

Eleventh Problem

> **999b24-7** Furthermore as regards principles one might also raise this difficulty. For if principles are one in kind, nothing will be one in number, neither unity itself nor being. And how will there be knowledge, if there will not be something which is one over the many?

This is the eleventh in the series of problems, but it comes now as the tenth exercise due to the fact that the sixth problem is left out entirely.[241] With this Aristotle enquires if principles are determinate in kind (*eidei*) or in number. For he says that [A] if they are determinate in kind, nothing of what is here below will be one in number, but will be one in kind, since all things here below have come to be according to these [principles]. Furthermore, nor will there be a science of things. For that which is known in relation to many things of like form must be one in number, if we are to reach knowledge, and the many be such that each of them is one in number, so that they will be like each other in form. But none of this will be the case, if principles are not one in number, but are determinate in kind. The whole structure of the argument is this: if principles are not one in number, nor are their effects. But if nothing is numerically one, nor can what is one among things of like form be grasped (for this is numerically one),[242] unless we seek a one superior to this. And so on to infinity. But if it is not possible to grasp a one over things of like form, then there will be no science.

> **999b27-1000a4** But if however there is something one in number and each principle is one (up to)[243] [, and if principles are not different for different things in the way perceptibles are (for example this syllable is the same in kind and the principles are the same in kind, and the principles themselves are different in number): if it is not thus, if the principles of beings are one in

number, there will be nothing else besides the elements. For there is no difference in saying 'the numerically one' and 'the particular'. For we speak thus of the particular, as the numerically one, but of the universal as what is over these. So therefore if the elements of a word were to be determinate in number, it would be necessary that all writings be as many as the elements,] there not being two or more of them.

[B] But if, he says, principles are one, not in kind, but in number, there will be as many of their effects as they. For as in the case of the 24 letters [of the alphabet], as elements of the written word, if they are so many, not in kind but in number, they would have made just as many words as they are (for it would have been necessary to take them [just] once), just as stones, were they determinate in number, would necessarily have made up just as many buildings as they are. Thus also the principles of beings, if they are determinate in number, will make as many things as they are.

'If principles [seen] as elements!', you might say, 'but not principles as generative and productive'. For in the latter sense, although they be determinate in number, nothing prevents principles from generating infinite things. But neither is the argument concerning principles which are determinate in kind entirely sound in its development. For it does not follow that, if principles are determinate in kind, so also are their effects. For matter cuts off and fragments what
41,1 is proper to immaterial and intellectual forms. But this has already been discussed above,[244] where we show that intellectual principles are determinate both in kind and number, but that the principles superior to them transcend this double [determination].

Twelfth Problem

1000a5-7 One difficulty no lesser than any other has been left aside both by those of today and their predecessors, whether there are the same or different principles of destructible and indestructible things.

This is the twelfth in the series of problems.[245] For in it Aristotle raises the difficulty whether there are the same causes for eternal and for destructible things, or there are different causes for the former, different causes for the latter. For if there are the same causes, how do they make, on the one hand, eternal things, on the other hand, destructible things? But if there are different causes, it is clear that the causes of eternal things are indestructible, yet some explanation must be given if it is held that the causes of destructible things are indestructible. For if [their causes] are indestructible, how is it that these make destructible[246] things, and not eternal things,[247] just as do the causes prior to them? But if the causes are destructible, will it not be that once the causes are destroyed, all of coming-to-be will collapse and come to an end?[248] The matter is therefore worthy of examination on the part of philosophers.[249]

But why does he say that this was left uninvestigated by all of his predecessors, when Plato, after the making of eternal things, clearly shows the demiurge as saying to the encosmic gods: '[there are] still three remaining genera of mortal things left ungenerated. The heavens will be incomplete if these do not come to be, for it will not contain in itself all of the genera of animals. But it should have these, if it is to be really complete. But generated by me and sharing in life, these would equal the gods. Therefore, so that all be mortal and immortal, as it should be,[250] turn yourselves following nature to the making of animals' (*Timaeus* 41B7-C4), and so on, where he teaches about the coming-to-be of animals, about nutrition, growth and decline?

[A] Plato therefore clearly says that the very first are causes of everything, that the [gods] in the heavens are in charge of coming-to-be, and that mortal things, were they not to have proximate causes which are destructible and changing, would not be mortal, having as they would simply an intelligible principle, as do those things which are eternal.

> **1000a9-14** Therefore those associated with Hesiod and all such theologians cared only for what was plausible for them, caring little for us. For making the principles gods and generated from gods, they say that those which did not taste of nectar and ambrosia became mortal, clearly using these words as if they were known to themselves.

Indeed, for what else than gods would one hold the principles to be? For nothing is superior to a principle, or to god and gods. In consequence the theologians must say that, and must generate all things from the gods: intellect, souls, the natures of bodies, both eternal and destructible things. Since, therefore, among the things generated, [i] some of them are united to the gods, [ii] some are such as never to be connected immediately to them, [iii] and some are related to them sometimes in the latter way, sometimes in the former (it is among these that we [humans] are ranked), the theologians say that some [i], on the one hand, – those which always look towards the gods – are nourished thence on ambrosia and nectar (ambrosia allows them to 42,1
be separate from coming-to-be, the place of blood and blood-filled impurity,[251] whereas nectar preserves from the spell (*thelgesthai*) that can come with caring (*promêtheian*) for the lowest things, allowing pure and unswerving care (*pronoein*) of the whole); [iii] some, on the other hand – those which sometimes reach to the gods, but are also such as to depart from there – these, the theologians say, when

they speak truth and are true to their oath (that is, when they are turned to real being and honour the divine), share in ambrosia and nectar, but are deprived of nourishment from the intelligible when they lie and break their oaths (that is, when they turn away towards non-being and generation and have contempt for the divine); and [ii] some, finally – those which are mortal and corruptible – never share in these foods. I, therefore, for my part cannot even conceive of how one might reach a truer arrangement of these matters, either as regards principles, or as regards those things which come to be from them. But if what is said by the theologians is not clear to everyone, it is perhaps because they practise lack of clarity (*asapheia*),[252] given how estranged is the multitude in relation to divine doctrines, perhaps also because the inspired way they speak[253] is such as to provide a teaching which is more enigmatic than discursive.

> **1000a14-18** Although what they said about the feeding of these causes is beyond us. For if they partake of these [foods] for pleasure, then nectar and ambrosia are not at all causes of [their] being, but if they are, then how would they be eternal if they need food?

These questions are crudely put. One should say in response that the illuminations and provisions of goods coming from superior causes are said to be 'foods', but that these also benefit to a greater degree those which already have the good in their own constitution, and give a share in a more perfect eternity to those which are eternal in their own substance. So that it is for the sake both of well-being (*euphrosunê*) and of being that they partake of them.

> **1000a18-19** But it is not worth devoting serious study to those who practice mythical deceptions.

We have not heard Plato saying these things about the theologians, but he deems it right to follow them in accounts concerning the gods, even if they speak without use of what is plausible and of demonstrations,[254] since they have an affinity to the divine surpassing other men, for which reason he is in the habit of addressing them as 'divine' or 'most divine'.[255]

> **1000a19-22** But from those who speak using demonstrations we need to learn by asking them why it is that those things which come from the same are, some of them, eternal in nature, others being destroyed.

Now it has been said[256] that this is the function of superabundance (*periousia*) of the principles, that the production of things does not cease until mortal things come to be in addition to eternal things. It has also been said[257] that the universe could not have been complete if it did not contain in itself all animals and indeed that, in producing (with the one demiurge) generated animals, the gods in the heavens necessarily make them mortal. For that which is generated from moving causes cannot remain forever.[258]

In addition to this, the position of Empedocles is preferred by Aristotle (1000a25) to others in this matter as not being at odds with himself at least in this much, that producing the same causes of all, 43,1 he makes all destructible and not some eternal and some destructible (1000b18-20). Aristotle again accuses Empedocles of contradicting his own hypotheses, for he says (1000b9-12) that Empedocles uses two causes, one being Love, a productive cause, the other being Strife, a destructive cause, but he destroys many things through Love and generates many things by Strife. Furthermore, in saying that they predominate in turn, Empedocles does not give the cause for this alternation of principles.

We, however, say that Empedocles does not destroy everything (for he is not in need of such self-consistency (*sumphônia*)), but like the other Pythagoreans, he also knows that there are intelligible substances and perceptible substances. Nor does he posit Strife as a destructive cause and Love as productive cause only. Nor is he silent about the reason for the seeming alternating predominance of the causes. But being a Pythagorean and Orphic, Empedocles first posits, as coming after the first principle of all things (about which neither he nor Parmenides nor Pythagoras permitted themselves to expatiate much), these two principles, Love and Strife, which the Pythagoreans called monad and dyad, calling the dyad indeterminate on account of its power which flows out in everything. From these principles emerge the intelligible and the perceptible worlds. Now in the intelligible world, called a 'sphere' in the poem, Love predominates by reason of the union of immaterial and divine substances, so that the whole seems an offspring of Love; but Strife predominates in the perceptible world. The cause of each of the principles predominating in each world (although both are everywhere) is the perfection appropriate to these that comes from the One that transcends both [principles]. This perfection Empedocles calls the 'perfection of time',[259] a 'perfection' proceeding from that which protects and holds together all things, 'time' being what Orpheus too called the First.[260] But Empedocles called the First 'oath' (*horkos*),[261] as the barrier (*erkos*) and protection of the others. And the perfection going out from it to the two principles of all things, giving measure to their powers and predominance, this he called the 'perfection of time', as if he were speaking of the perfection coming from time.

1000b14 But when great Strife was nourished in the limbs.

He speaks of this as 'great', as does Orpheus of chaos: 'and a great chasm, immense, to this side and to that',[262] and as Pythagoras speaks of the dyad as indefinite. Empedocles gives it limbs and growth, since it is the cause for all things of going out, coming-to-be, plurality, multiplication and all growth.

1000b15 And sprang to honours with the perfection of time.

It enjoyed[263] honours and kingship corresponding to the perfection which time brought it, for the very First also measures the infinite power of these two principles.

1000b15a Which alternately was fixed for them by a broad oath.

44,1 This perfection they received from that which protects wholes, the very first principle of all things, a perfection called 'broad' as uncircumscribed by all things and as circumscribing and measuring all things. 'Alternately': not that the One that transcends the two principles confers perfection sometimes on the one, sometimes on the other, but that it confers on both the perfection proper to each and appropriate to both. But if what follows [i.e. Time] is also 'alternately' in that it is always with them and is in no way missing from anything which is, and certainly not from the two principles that come after it, this also would be rightly said. For the goods given by the very first cause of all things are present to them all.

But how could Empedocles make the intelligible god lacking in any sense (*aphronestaton*, 1000b4), a god in which is truth itself? Because, Aristotle says (1000b4-6), there is no Strife in him, but for him [Empedocles] like is known by like. Yet there *is* also Strife in him, through the unification in him and through sameness. For which reason Parmenides also said that being is one. Then again, if *we* 'see earth with earth'[264], why should it be thus for the divine? And how can he [Aristotle] fittingly say that [the divine does][265] not know everything? For thus would his first god, thinking only himself, have less than we who can turn our souls' eye to all things. But let what has been said suffice as regards these matters.

[B] In what follows, Aristotle raises difficulties for those who posit different principles for eternal and for perishable things. For if, he says, the principles are <destructible>,[266] these philosophers destroy the principles of destructible things, and thus to infinity. For it is clear that they are destroyed into that from which they came, so that there will be principles of principles, and so on to infinity. Furthermore, how will all coming-to-be not collapse if its principles are destroyed? This is all well said and true to Plato (*platônikôs*).

If, however, the principles are indestructible, why is it that they also do not themselves produce eternal things, as do the principles prior to them? But it has been said that the principles are indestructible, but not immobile: and this is what he intimates with a 'long account'(1000b32). He said that his predecessors 'nibbled at' (1001a2) the difficulty, in the sense that in touching it, they run away from it like mice which through greed approach food, but do not think it safe to dally around it: thus are these philosophers like those who nibble and take a taste of something.

Thirteenth Problem

> **1001a4-9** But the most difficult of all things to study and most necessary for knowing the truth is the question if being and unity are substances of beings, and if each of these is unity or being, without being something else; or if one should enquire what being and unity are, as belonging to another underlying nature. For some think things are such as to be in the former way, others in the latter.

This is the thirteenth in the presentation of the problems, the twelfth here, because the sixth has been entirely left aside, as we have often indicated. In the first part[267] the positions of the Pythagoreans and of Plato have been stated, and that in these matters Empedocles is not in disagreement with them, whereas the natural philosophers from Ionia think perhaps that unity and being are water, or fire, or suchlike. But yet Aristotle himself, even <if>[268] he also advocates an intelligible and unmoved substance and rejects those who say that there are only bodies, clearly does not separate unity and being in things, but says they are the same by [their] substrate, differing (if at all) from each other conceptually (*epinoiai*), each and both of them signifying this particular thing (such as intellect, or soul, or universe, or sun), and does not accept that there is unity itself or being, as being anything other than the things that are. Perhaps it will be possible, when the occasion arises, to raise difficulties for him on these points, but now, for the moment, let us consider his arguments for both [of the opposing] positions.

> **1001a19-24** But it happens that if one will not posit unity and being as being substance, then none of the other universals will be this (for these, most of all, are universals, and if there is not a unity itself, nor being itself, there will hardly be the other universals besides what are called particulars).

[A] The argument has much cogency and corresponds to what seems to be the case. For if there are universals, much more so is there what is most universal. And if there is what is most universal, [then] there is unity and being which transcend all things, for these are what is most universal. But if these are not, then there are no universals. From which it follows that neither is there science nor things which in their nature are like in form to each other, but it is by chance, precariously and arbitrarily that these things are like in form to some and unlike others. For if one form does not pre-exist, in relation to which that which shares in it is of like form and that which does not share is <not>[269] of like form, I do not see what cause there will be of these which is productive, protective and permanent.

Not only this, but if one were to examine the argument, one would see first that forms and universals are destroyed by the hypothesis, but also, and not less, the determinate perceptibles among beings. For were there not to be Unity itself, transcending all things, there could be neither anything universal nor anything individual: as regards the former, since [its] nature is 'one and many',[270] what is one only must be ranked as prior; and, as regards the latter, because the composition [of individuals] is a 'many which is one', and at the same time, since they <are one>[271] by participation, if we are not to place as prior the unparticipated one, there will not be a cause of unity in beings, just as those thinkers who say that Being is not a principle of all would leave no cause of being to those things which in whatever way are. For as the Good is the single principle of goods for all, which itself is neither the good of something nor anything but the Good; and as the self-moved is the cause of movement for all, and for it Movement itself (*autokinêsis*), not the movement of this thing, nor anything else but movement, but that which we say is precisely Movement itself; and as all the things which know have Knowledge itself (*autognôsis*) as a principle: in this way, then, beings as beings proceed from being, which is not this [particular] being, but Being itself, and things which are unified receive unity from the One which transcends all things.

1001a24-6 Furthermore,[272] if unity is not substance, it is clear that there would not be number as a nature separate from beings.

It follows necessarily that there will be no substantial number, if unity is not, not however for the reason which he wishes to adduce (which is verbal (*logikê*)). Now he falsely takes it that form-number, substantial number,[273] is made up of monads, although if one were to agree that this number comes from monads, it would not be from
46,1 monads like in form to the unity at issue in the argument. In what way then is there truth in what is said, that everywhere number

proceeds according to resemblance to unity? Is it not that if there is a unity transcending all things, there will also be number, not number as in what is numbered nor monadic number (for these come last and come to be with another unity), but number itself, being an intellectual form and the primary and intelligible substance. But if there were not unity itself, there being no principle, neither would what comes from the principle come to be.

1001a27-8 But if there is a unity itself and being, unity and being must be their substance.

This also is well said. He uses [the term] 'substance' in a more common sense in connection with unity, since being only unity, it is clearly above substance, if, that is, substance is in need of being one, whereas unity does not need substance, lest instead of being unity it becomes the unified.

1001a28-9 For these are not something other, [that of which] they are predicated <universally>,[274] but are these themselves.

For they are not predicated universally of other things as being other than them, as are species and genera (for species are predicated of individuals, being other than them, in the same way as genera are predicated of species and individuals, being other than them), but their substance *is* unity itself and being itself. Thus as we speak of Socrates, not predicating something of something else, but saying what precisely he is himself, in like manner we do not assign unity and being as other in relation to what is other than them, but we name them in such a way as to express them as themselves. What Aristotle says is true, but one should add that even if these are not said of something other, they are the causes of being and unity for all things.

1001a29-31 But if indeed there will be being itself and unity itself, there is much difficulty as to how there will be something else besides these, I mean how there will be more beings than one.

[B] He begins the argument attacking the opposed position, saying that the account of the existence of beings is most difficult for those who posit Being itself. Now that the question is profound as to how from a unity which has no duality, no trace of plurality in it, no otherness, all things come to be, this is made clear by Plotinus, Porphyry, Iamblichus and all those who have treated of this problem with a greater degree of knowledge.[275] Nevertheless we will undertake to resolve the difficulties developed by Aristotle, giving as an explanation the cause, most generative in [its] simplicity, belonging to the One.

> **1001a31-b1** For the other than being is not, so that it will turn out to be necessary, as Parmenides' account has it, that all beings be one and this is being.

We have said a number of times in what way Parmenides said that being is one: it being his view that all the intelligible is supremely unified, he said that being is one and that which is besides being[276] is
30 not, which is what he meant by the perceptible [world]. And what is not is nothing, because not equal in rank to being, nor is it something [as if additional] to it [being], nor does [being] become more if [what is not] exists.[277] For as Socrates remains one, even if there are many images of him, thus does the intelligible remain one, even if this world exists, which is nothing to the intelligible, apart from the world depending on it and taking its being and perfection from it.

But it would not be reasonable to agree with Aristotle when he
35 says that there would not be anything other than being. Firstly, the hypothesis that is criticised by him here does not take being in the sense of the whole intelligible reality, but in the sense of what is highest in it, which is neither intellect, nor life, in some sense, but Being itself, from which those which have it and are its proximate
47,1 products are different, while being yet beings. Then again, if Aristotle himself agreed that perceptibles are not beings, he should have accepted the premise mentioned, that what is other than being is not. But since he subscribes to perceptibles, more than to anything else, as beings, how could he properly assent to this here? We therefore accept neither [the alleged] Parmenidean
5 being nor non-being. Thus, in <not>[278] agreeing to the latter, we reject the present critique.

> **1001b1-6** There is difficulty both ways. For were unity not to be a substance or were there to be a one itself, number could not be substance. We have said earlier why, were it not to be [substance]. Were it to be, there is the same difficulty as that in regard to being. For from whence is there also[279] to be another one besides the one itself? It would necessarily be nothing,[280] but all beings are either one or many of which each is one.

Now it has been said, and well said, that if there is not the One, there will be no number. For if, he says, the principle is destroyed, neither would it ever come from something, nor something else from it. But
10 why, if it exists, will there not be [number]? Is it because of the beginning of the first hypothesis in the *Parmenides*, 'if the one is, there will not be many' (137C4)? But there [Parmenides][281] elimi-

nates, not the many that proceed from the One, but the many that are in the One. For if it is in reality purely one, it would not be this itself and also many. But this does not prevent it from being the cause of many intelligible, intellectual and cosmic numbers. But Aristotle understands that what is besides the One is nothing. For all beings are either one or constituted of monads, so that if there is some one coming from the One, it will be other than the One. But being other than the one is nothing. But nothing is not, since all being is 'either one or many of which each is one'.

But we deny that what comes from the One is already nothing. For if it is not the One, it is not immediately not one, or nothing, but one in a secondary sense or again in a tertiary sense. For when the last of the last, perceptibles which are destructible and enmattered, share in the one, would then intelligibles and divine things be deprived of being one? It suffices therefore that the generative cause in the One produce from itself both numbers and all things. And there is no need to seek a substrate out of which beings come to be. For it is nature that is the first to have need of it [a substrate].

1001b7-8 Furthermore if the one itself is indivisible, it would be nothing according to Zeno's axiom.

Aristotle says that Zeno's argument is crude, since he destroys the nature of the one, of the point and of all that which is indivisible, on the grounds that neither when they are added nor when they are subtracted do they change the quantity of substrates. For he who says this, Aristotle says, thinks first that all being is magnitude and this a solid magnitude, for surfaces and lines, when they are added or subtracted, do not diminish or augment the substrate in every possible way, but augment it if put together at their limits, but otherwise not.[282] Secondly, it is argued that even if the monad does not make a quantity greater, it makes it more in number.

But we do not think that Zeno argues seriously here, as he is producing argumentative exercises (*gumnastikous logous*) [directed] against the common folk: unless perhaps he hints in this that indivisibles do not belong to the nature of divisibles, but are superior to them and limit and define them in another way.

1001b17-18 But how will magnitude come from a one of this 48,1
sort or from many such ones?

Earlier it seemed to Aristotle a surprising thing that even numbers should be able to come to exist from a one of this sort. But now he says it is yet more strange to derive magnitude from it. This is why he added:

1001b18-19 For it is the same as saying that the line comes from points.

But we say that there is, on the one hand, magnitude itself, which has no parts, is divine, intelligible, established[283] among the first of beings, which manifests itself from there[284] with all true beings, and, on the other hand, perceptible magnitude which is conferred on matter by nature which is empowered and guided by the demiurgic gods, of which the father[285] is the one universal demiurge, offspring of the Good,[286] which is to say the One, in imitation of which all things confer unity on those things coming after them, such that perceptible magnitude is also unified by it. How, therefore, can it be the same to say that the line is constituted out of points (which has been shown to be quite impossible) and that the One produces the many sorts of magnitudes? For there is nothing common to these, unless it be that, as all other things wrest off a trace and last image of the One, so also the point in its indivisibility would seem to imitate the One. But it is imperfect and does not exist of itself (*anupostatos*), what it is it has from another, it is sterile, lifeless, irrational, whereas the One is the cause of all perfection and of what really is, remains in itself, is most generative of all things, the principle of life and thought, and superior to all that of which it is the principle.

1001b19-25 But if indeed one were to take it that it happens thus, as some state it, that number comes from the one itself and from another which is not one, one must all the same investigate why and how the product will sometimes be number, sometimes magnitude, if indeed what is not one be inequality and the same nature. For it is not clear how magnitudes would somehow come either from the one and this nature, or from a number and this nature.

Aristotle raises difficulties for Plato, as to how he here makes number, and there magnitude, from the One and the Indefinite Dyad which he calls 'not one' and 'inequality'. For it is not clear how magnitude comes to be from the [Indefinite] Dyad, if number is co-ordinated with the Indefinite Dyad rather than with the One. To this we say that the intelligible monad and dyad generate number [in the reality] beyond, whereas what corresponds to them in our [world] produces perceptible magnitude. Form in our [world] is, as it were, a one and a monad, and matter is the ultimate imitation of the Indefinite Dyad, from which comes magnitude here below. But if you wish to rank nature, on account of its power of generation, with the one and monad, then you will consider form as relating to number. Consequently, one might say that in one sense magnitude comes to be from the monad and the Indefinite Dyad, the monad not providing itself to the composition of the magnitude, in another sense it comes from number and the Indefinite Dyad which are as if elements of it.

Chapter 5
Seventeenth Problem[287]

> **1001b26-8** Among these there is a difficulty as to whether numbers, bodies, surfaces and points are substances or not.

This is the last, the seventeenth, in the presentation of the problems, but it is here the thirteenth, due, I think, to its affinity to what has been the object of exercise before (49). For having enquired there[288] whether unity and being, as what seem to be the principles of intelligible substances, are substance, he now wants to examine whether mathematical knowledge has to do with substance. But it has been said in the first part[289] which of these are substances and which one must think of as accidents. Thus let us turn now to the arguments.

> **1001b28** For if they are not, it escapes us as to what is being and what are the substances of beings.

[A] If, he says, one denies that these are substances, we will not yet know what is substance. For whiteness or whitening are not substance, but the one is an affect of substance, the other a movement. Nor indeed is 'to be on the right', or whatever is said to be a relation, substance. Nor is health, nor are the accounts (*logoi*) concerning these, but one is a disposition of substance, the other sounds which signify. Thus all these things are in a substrate and none exists of itself. But as for water and fire and air and earth: these are not substances with respect to their being qualities, but with respect to the body without quality which we say persists through the various changes – perhaps this might have something of a substance (*ousiôdes*) in it. Except that of this it would seem that what is more of a substance is the surface, and of the surface the line, and of the line the point, and of the point the monad. For everywhere what limits and defines is more substance than what is limited. And prior in substance would be that which removes but is not removed.[290] Thus the monad defines all things, the point extension, the line both surfaces and solids, the surface bodies. And if there is no monad, there is nothing else of what comes after it, or if [there is] not the point, then [there are no] extended things. But if there is no body, what is prior to it is not prevented from existing. Thus either beings are bodies, according to the philosophers of nature, and the principles of bodies are substances, or they are numbers, according to the Pythagoreans, and the principle of numbers would be substance as would be what is related to the monad.

But we say that neither of these points argued in this way has true strength. For it is not always the case that what limits is, to a greater degree, than what is limited, but it is so when a limit, transcending the nature of something, is conferred on it, such as skill or nature conferred on fabricated or natural bodies. But it is not so when the limit [actually] contributes itself to the limiting of what is defined. Nor is it always the case that something, if it removes [something] whilst it is not removed [if the other is removed], is more a substance than that of which the opposite holds. Since, were this the case, then stones would be, to a greater degree, than the house, and pieces of wood than the ship, and in general the matter than the things which are formed. But it is the case when that which removes is not a part of what is removed, nor an element, nor a limit, nor anything which wholly contributes itself to the constitution of the other.[291] Therefore, true genera are prior to forms and substance is prior to accidents, whereas surface is prior to body as more immaterial,[292] but not prior as substance to a greater degree. It is clear that the arguments
50,1 concern surface as that which exists *within* bodies.

1002a15-18 But indeed if this be agreed, that lengths and points are substance to a greater degree than bodies, but we do not see of what bodies these would be (for it is not possible that it be in perceptibles), then there would be no substance.

[B] If these are substances to a greater degree than bodies, but there are not these things (for they are not in perceptible bodies, nor in others, but they must be in bodies), how could there be a substance? But it should be said that they are both in perceptible bodies, in a natural and enmattered way (thus they do not preserve the nature which they announce, no more than any other enmattered form); and also indeed in mathematical body, which if not perceptible is indeed known in imagination (*phantaston*); and, furthermore, in the reasonings of opinion (*doxastikois logois*) and in intelligible forms. And yet as these exist in these many ways, nothing prevents there being other substances corresponding to each of the ranks of beings.

1002a18-20 Furthermore, these all seem to be divisions of body, the one into breadth, the other into depth, the other into length.

Divisions are not substances; the division of magnitude in depth is surface, in breadth is the line, in length is the point;[293] thus these will not be substances. But it should be said as regards the major premise that it may be that the divisions, for example, of monadic numbers, divisions that are within bodies, are not substances. Yet as regards the first otherness, which distinguishes everything, and in general the divine genera which are responsible for division: why should these not contribute to all things the division established according to forms and substance?

1002a20-4 In addition, any or no shape is present in the same way in the solid. So that if Hermes is not in the stone, nor is the half of a cube in the cube as something delimited. Therefore neither is the surface (for if anything were, it would be the surface that delimits the half).

<This is>[294] the argument. Neither are the shapes that could come to be in the stone in it, nor are the divisions which are not yet in act. If <such>[295] were [the case for] the surface, neither would any other [division be in it]. But to this it might be said that it is not[296] necessary that if there is what is in act, that there be what is in potentiality, or if the latter is not yet, that the former not be either. But nothing prevents there being something [a division] in act, such as the act which divides the cube. For if divisions are infinite in potentiality, nothing stops our seeing as in act <the divisions>[297] of what is limited.

1002a28-9 For in addition to what has been said, what also concerns coming-to-be and destruction makes no sense.

The argument goes like this. All substance which sometimes is, and sometimes is not, is brought to being through coming-to-be and to not being through destruction; but these limits of magnitudes[298] either are or are not in a non-temporal way; so they are not substances. And the argument seems to have some truth, since indeed all substance which comes to be and is destroyed requires a material cause. But these [mathematical objects] are not enmattered things. An indication that the point before was not and, as if hidden, later becomes manifest, is the fact that when a line is divided, this makes two points come to be. 'For the one point', he says, 'being indivisible is not divided into two' (1002b3-4). But it is clear that points are not before, in a non-temporal way, and receive existence out of the division, without coming-to-be.

1002b5-6 It is similar with regard to the present moment in time.

He also says that the following is an indication that there are some things which, not being before, appear outside coming-to-be and disappear without destruction, none of which are substance: the present moment in temporal division, to which also corresponds the point. None of these are therefore substance. To which you might respond that even if these are not substance, this does not prevent them from being beings, for being is [said] in many ways.

Chapter 6

> **1002b12-13** But in general one might raise the difficulty as to why, besides perceptibles and intermediates, one must search also for other such[299] things, such as we who posit the Forms.

This problem he did not present specifically in the first part [ch. 1], but it seems to be part of the fifth problem (for he raises difficulties there concerning the Forms), or of the ninth (for there also he raised difficulties about the Forms). The whole argument runs as follows. If, besides perceptibles and mathematical objects we posit Forms for this reason, that since each of the perceptibles is many (such as many horses and cows and men), as are mathematical objects (for there are many triangles and circles and spheres and pyramids), and we must not stop at plurality, but must go to the monads of each (the one man and the one triangle and the one circle), [then] each of the Forms must be one in number. Yet it is not possible for the principles to be one in number, as has been shown in the tenth problem.[300] So there cannot be Forms.

To which you might respond that perhaps neither have material causes been shown as not all being able to be determinate in number: Empedocles says <, for example, that the roots of all things are four>.[301] Yet if therefore, as regards the principles that are *in* the products, it appears as impossible that the principles have a numerical limit, then these criticisms will have no hold with regard also to the generative and superior causes. For thus neither would there be the 'separate' numbers (in his words) if they were <not>[302] determinate [in number]. And it is a false assumption to think that it is only for this reason that those who love to know the truth[303] have recourse to the Forms. However, it is correct, and we agree, that all that which the Forms have 'is substance and in no respect by accident' (1002b29-30).

Fifteenth Problem

> **1002b32-4** Related to these points is the difficulty as to whether the elements are in potentiality or in some other way.

This is also the fifteenth in the presentation of the problems. Aristotle asks here if the principles are in potentiality or in actuality. For
52,1 the [words] 'or in some other way' signify what is in actuality, but I think he conceals this term on purpose, so as not to remove what is difficult in the problem. For who would not have said that the perfect is more a principle than the imperfect? For the same reason he also did not speak of causes or principles, but of elements, in which in

reality there is much imperfection. As to what the case is, we have said this in the first [part].

1002b34-1003a1 For if it is somehow in another way, then there will be something else prior to the principles.

[A] It is as if he said: if one will posit the principles as in actuality, since potentiality is prior to all actuality, then there will be something prior to the principles. But he has suppressed here also the term 'actuality', with the fact that the argument is [purely] verbal (*logikon*). For if potentiality is prior to actuality for each of the things that come to be, this is not so for things which are wholes or divine, as he himself has shown us elsewhere.[304]

1003a2 The potential need not all be in this way.

All that which is in potentiality, he says, need not come to be in actuality, but what now is in actuality was before in potentiality.

1003a2-3 But if the elements are in potentiality, then it may be that there will be nothing of what is.

[B] He argues for the opposite position, that one should not posit the elements as in potentiality. The argument runs like this: only what is possible comes to be, for it is not the case for what cannot be, or already is; what comes to be is not yet; it is possible for what is not yet not to come to be; so that, if the principles are in potentiality, there may be nothing of the beings which come from them. He explained the minor premise thus: 'nothing of what cannot be comes to be', which is equivalent to only what is possible coming to be. The middle premise is this: 'for what is not is what becomes', for what becomes is what is not, hence one would not speak of what is already in actuality as coming to be. However, it is possible for what is not yet not to come to be, as he brings out from these premises, since this is what it is to be possible, to be of a nature such as to progress to actuality, but yet this might not occur in every case. But all this [argumentation] is very strong, if one were to say that all principles are in potentiality and none in actuality.

Fourteenth Problem

1003a5-7 Therefore it is necessary to raise these difficulties concerning principles, and whether they are universals or what we call particulars.

This is the fourteenth [problem] of those mentioned in the first [part],

but it is much the same as the tenth. For here also he enquires if the principles are universal or individual, which is much the same as asking if they are determinate in kind or in number. But he says [A] that if we were to posit the principles as universal, we remove what is substantial in them. For universals are not substances. For if they are substances, then Socrates is at least three substances: the individual, substance by species, and substance by genus. But [B] if the
53,1 principles are not universal, either they are unknowable, or there is something prior to them. But each of these options makes no sense. For if they are not known <then nothing is known; if>[305] the principles <will be known>, they will be known through universals. For it is only in this way that individuals are knowable. There will therefore be something prior to them which is comprehensive to a greater degree than they.

But it has often be said that, beyond both the individuals and universal reason-principles, both those which are productive and those which are known, there are what are principles in the primary sense. Nor has what Aristotle says here any force. For not all that is universal is without substance, for then there would be nothing universal which is eternal and productive of particulars. Nor do we know the particular things (*ta tode ti*) through universals, [by] having recourse to a universal produced *a posteriori* (*husterogenê*) which is nothing.[306] Nor would he himself say that anything universal transcends the immobile genera.[307]

Conclusion

But as the commentator Alexander has also indicated,[308] the philosopher [Aristotle] has gone through nearly all these matters using dialectical arguments. (The topic that comes sixteenth in the presentation of the problems, whether the principles are mobile or immobile, is not the object of exercise here.) Having presented thus seventeen problems in the order we have explained, Aristotle did not elaborate arguments for opposite positions with regard to the sixth and sixteenth. As regards the sixth: because, in addition to the reasons already given,[309] it is also anticipated.[310] For in enquiring if the first philosopher knows not only substances but also their essential accidents, he was enquiring also in a way if the philosopher will deal with the same and similar and equal and their opposites, for these would be the essential accidents of being as being. And as regards the sixteenth problem: since he everywhere admits that, among beings, some are destructible and some are eternal, it would necessarily follow for him that, among principles, some are moved and some are immobile. But it is clear of which beings which principles, for he has shown adequately that it is not the same principles of both kinds of beings. Adding to the remaining fifteen the fourteenth in the order in which the problems are elaborated, that which objects to the Forms, Aristotle put forth all sixteen problems for the purpose of dialectical exercise (*gumnasian ... logikên*), of which some he will arbitrate in book 4, some in books 7, 8, 9, 10, but most in book 12. He will attempt to argue against the Forms and numbers in the last two books, books 13 and 14.

Syrianus, Son of Philoxenus 54,1
Introduction to Book 4 of Aristotle's Treatise *Metaphysics*

This book is expository and is not devoted to raising problems as is the preceding book.[311] Here, rather, Aristotle solves many of [these] problems. For he teaches here that it belongs to one <science to know> the causes of being as being (*tou ontos hêi on*);[312] that all substances and the essential properties of these are examined by this science; that he who practises first philosophy (*ho prôtos philosophos*) also deals with the principles of demonstration. He will also discuss the axiom of contradiction, as being the most primary and most comprehensive of all that which is [found] in the things demonstrated in the sciences, whether it indeed be demonstrable or possesses some trustworthiness (*pistis*) superior to demonstration. He will attempt to convey <these matters>[313] in this book, which, given the adequate explanations provided by the most industrious Alexander,[314] we will not interpret in its entirety. But we will try to examine (*basanizein*) the part where we feel Aristotle says something which causes difficulty <and merits>[315] explanation, summarising (*paraphrazontes*) all the rest so as to respect the continuity of the treatise.

Chapter 1[316]
The Scope of First Philosophy

> **1003a21-3** There is a science which examines being as being and its essential properties. This science is not the same as any of the others.[317]

The philosopher has also noted that, if there is this science, it is not the same as any of the other sciences when he says (1003a23-5) 'for no other science examines being as being in general, but cutting off a portion of it, considers its[318] accident'.[319] The inductive proof (*epagogikê pistis*)[320] of this is taken from the mathematical sciences, from geometry, astronomy, arithmetic.

But why must there be such a science of all being as being? Is it not especially – as is often his practice, he first states what he is going to show and then will show it clearly, shortly afterwards – because he takes it [i] that 'being', if indeed it is not said of all things univocally, nor indeed equivocally, is like terms which are said by derivation from one thing and by reference to one thing, and, he adds [ii], of these terms there is one science? Then again you may reach this conclusion on the basis of the arguments he has given previously. For if what is knowledge to the highest degree concerns being in the highest degree, and the truth and clarity of the knowledge corresponds in degree to the high value of its object, as has been said in Book 2 (993b27-30), then there would be a science of being as being, and not only this, but this science [must be] the finest and the best of the sciences.

For there is nothing more worthy (*semnoteron*) than being. For either [i] there is something which is, and is nothing else (such as intellect, or soul, or the heavens, or the world), uniquely existing as what is, as Being itself (*autoon*), clearly more worthy than all else, itself not needing to be a world, or the heavens, or soul, whereas all the rest receive their being from it. Or else [ii] being must be something belonging to them, but then it will be of more value, as being, than in respect to anything else. For everywhere being and substance have precedence, and substance is the hearth (*hestia*) of things. How then could this remain unknown, being the best of whatever is in any respect, or philosophy, true philosophy that is, have to do with anything else? For it would be quite right to hold that the first of sciences concerns what is first, the best science concerns what is best, and the whole concerns the whole.

But what then: is the Good not that which is even beyond being?[321] Why then is the first of sciences not of the Good, as Plato seems to be saying in the *Republic* [505A2-4]: 'since the Form of the Good is the highest object of knowledge, as you have often heard, by which all that is just and in general serves some purpose becomes useful and beneficial'? But is it not especially because the good, according to Aristotle, is not different from being? For the absolutely primary, for him, is both being and the good, as he clearly teaches in book 14 (4, 1091b16-20), which is also why, in the *Physics* (2.7, 198a25-6), he identifies the goal with the form, for the form derives from being and the goal from the good. Thus in this way also, according to Aristotle himself, the very first and best of the sciences will concern itself with being as being. Furthermore, for those[322] for whom the Good transcends being, it is unknowable and above all science, as Plato clearly proclaims in the *Parmenides* [142A4-5], whereas the passage cited from the *Republic* speaks rather of the Form of the Good which is seen *in* and *with* beings. Thus again the very first of the sciences is ranked with (*suzugos*) being as being, according also to the more contemplative of the philosophers. That it is not otherwise is sufficiently explained for us by Plato himself in the *Republic* in the division of the line [511B3-D5].

But if wisdom is the knowledge, not only of being as being, but also of its essential accidents, so that it would also know the good in being and have an altogether perfect knowledge of it, it would also be a demonstrative science, or rather it would be demonstrative science as a whole: not as a whole produced from what are rather parts, but as a whole prior to the parts.[323] For this reason this science also provides the principles of demonstration to the other sciences. Thus, ascending by means of demonstration, it considers the essential 56,1 properties of beings, through which it reaches up to the distinctive property (*idiotês*) of being. Remaining above after that, it proceeds according to the natural order, distinguishing by means of division and definition the substances of things, and, by means of demonstration, the essential properties.

Thus this science will also investigate the first and highest causes,[324] the causes not of what exists in a substrate, but of beings which can exist as being of themselves and in relation to themselves. Perhaps indeed the earlier philosophers, as they sought the elements of beings, wished to reach knowledge of such principles,[325] which are the principles, not of accidents (for the principle *per accidens* is the same as the principle of accidents),[326] but of essential [attributes] and of beings as able to exist of themselves. But let us for the moment leave aside the question as to whether these philosophers succeeded or not in their intention (*prothesis*): Aristotle says (1003a31) that it is necessary, for himself, to seek out causes of this sort.

Chapter 2
The Structured Multivocity of 'Being'

Now Aristotle has shown elsewhere[327] and takes it here that 'being' is not predicated univocally of all beings, nor that beings as beings are of equal rank (*isotima*). Neither indeed are beings, he says (1003a33-4), simply equivocal, but even if being is said indeed in many ways, it is still by reference to one thing, i.e. to the one nature of substance (*ousia*). For just as 'healthy' is not said univocally of exercise, of food, of colour and of the body, but is said by reference to and by derivation from one thing, health (the one as preserving health, the other as producing it, another as a sign of it, and body is said to be healthy as receiving health); and just as a slave[328] is said to be 'medical' as having learnt and as possessing medical skill, or as having a nature suitable for learning, or again something is called 'medical' in regard to what is achieved by medical skill, 'medical' being predicated in these cases neither univocally nor equivocally, but by reference to and by derivation from one thing, so it is in many other cases. For example, 'good': for god is a good, as is opportunity (*kairos*), and virtue is a good, as is form.[329] Thus it is also in the case of being. For being is also said in many ways, and it is not said univocally of things, but all are referred back to substance (*ousia*) and it is on account of substance that they are spoken of as beings.

For substance is properly, primarily and especially being. But we also speak of the affections (*pathê*) of substance, such as warmth, chill, <dryness,>[330] wetness, and all things of this kind, as beings. And indeed we speak of the processes of coming-to-be, considering them as certain ways leading to substance, as beings. And not only the processes of coming-to-be, but also the processes of destruction, since they are themselves the destruction of substances. And we speak of processes of maiming and becoming deaf as beings, because they are considered as supervening by privation (*sterêtikôs*) in relation to substance. We also call beings those [things] which produce substances, such as the activities of productive substances. Similarly, we speak of that which is generative,[331] such as that generative of animals we see[332] produced through some humid putrefaction: even of these things do we speak as belonging to beings. And if something be productive or generative of a quality, we deem it worthy itself of being called a being, since all accidents are said with reference to substance, even if this thing produces a relation only, since relations arise as accompanying substances. Thus we speak of all these as beings in reference to substance.

And what is surprising in this, when we even speak of negations of accidents and of negations of substances themselves as being? For even 'not white' and 'not man' in some sense 'are'. For here is manifest the marvellous and really miraculous (*daimonia*) power of substance. For if it gives being even to that which is as if opposed to it, why would it not fill all things with being and seek to give its support unstintingly[333] to all things? Thus, in the way that the Good is cause even of what are said to be evils, and the light cause of darkness, so also is substance a cause [of the fact that] non-being (*to mê on*) is, in a sense. If then we say that non-being 'is' non-being (1003b10), we speak the truth in predicating 'is' of it.

Furthermore, if all beings are said by reference to one thing, substance, what is surprising in our saying that there is one science of all being as being? For as there is one science of all that which is healthy, as said by reference to one thing, health, thus there is also one science of all beings. For it is not the case, for those objects to which one science is assigned, that they are in every respect totally univocal, but there is also one knowledge of these things which are said by reference to one thing. For if indeed these things are intermediary between equivocals and univocals, yet they tend more towards univocals. It is therefore clear that it is for one science to consider beings as beings.

However, it is not because this science considers all beings that it assigns itself equally to all. Rather, it will occupy itself with that which primarily is, that is in particular the substances on which depend the other things which are deemed worthy of being called beings. Thus we spoke correctly in our prefatory remarks, when we said that the philosopher wishes to possess the causes of substance.[334] And indeed three claims were made in the preface: that the wise man deals with all being as being; that he deals with the essential accidents in being; and that he considers the causes of substance.[335] Now of these three claims, we have already demonstrated the first, in finding that being is not said equivocally, but relates entirely to one nature, that of substance. As for the third claim, we showed it through saying that the wise man will be particularly concerned with substance, which he cannot grasp without examining all of its causes. But there remains to secure the second claim also, which we will do,[336] to the extent that it is not yet done, having taken up first in addition the following points.

That which underlies and is of the same kind, what he has here called a genus (1003b19), if it is perceptible (*aisthêton*), correlates to one of the senses,[337] as for example colours to vision and flavours to taste and noises to hearing. But if it is an object of science (*epistêton*), then again it is grasped by one science. For the grammarian will examine all sounds (*phônai*) and the geometer all figures and the musician all harmonies. In this way then the philosopher considers all substances, even if, of these substances, some are intelligible and 58,1
unmoved as well as eternal, others eternal but mobile, others subject to generation and destruction.

Nonetheless it would seem again that no one deals with *all* that comes under a category. For the geometer does not deal with all quality, nor the arithmetician[338] with all quantity, nor the musician with all relation, but each deals with one species of those things which come under one or other of the categories. This is because none of them deals with quality as such, or with quantity as such, or with whatever else is most generic, as such. For if he were to deal with such an object, he would absolutely have to work with the category as a whole. Thus, either we must give up the idea that there is a science dealing with substance as being substance, or, as long as we accept this, we will agree that it deals with *all* substance.

Yet, just as philosophy, as one and a whole, will deal with all substances, so also will its species (*eidê*) deal with the species of substances.[339] Therefore, first philosophy will deal with intelligible substance; philosophy which deals with the heavens will concern eternal substance which is in motion; and a philosophy will concern substance subject to generation and destruction. But it is also possible to make the more specialised sciences two, naming the one 'first philosophy', and the other 'physics', as concerning that which is in motion which is both destructible and eternal. For depending on the way in which we divide substances, thus correspondingly will we divide the kinds of knowledge of them.

Thus, another of the puzzles developed in book 3[340] is now demonstrated. For there (995b11-13) the difficulty for Aristotle was whether one science will deal with all substances, or different sciences with different substances. He has now shown (1003b19-22) that philosophy as a whole concerns all substances, but that the species of philosophy will each concern one [type of] substance.

The Essential Properties of Being
1. Unity and Being

1003b22-3 If being and unity are the same and one nature.

He wishes from this point to show that it belongs to philosophy, as one and as a whole, to consider also the essential accidents of substance, taking the demonstration in hand in the following way. Being and unity are one nature; consequently, if the philosopher examines being as being, then he will consider unity as unity. But indeed it is for the same person to consider opposites. So he will also consider multiplicity as opposite to unity. But if he examines unity and multiplicity, and if all contrariety comes under these (for the same, the similar and equal are, in some way, one; the different, dissimilar and unequal are a dyad and multiplicity), then he would consider all that is opposite in this way. But these are the essential properties of substance. For they are primarily present in substance, and in all substance. Thus it belongs to the philosopher also to know the essential properties present in substances.

Up to this point, given the way in which Aristotle has approached the argument, we accept more easily that the essential properties of being are opposites to each other; that all opposites come under unity and multiplicity; and that it is for one science to know these. And indeed what seems most forceful in this argument, that all opposites come
59,1 under unity and multiplicity, receives much plausibility (*paramuthia*) from what Aristotle himself said, but is expressed in a most outspoken way (*pleistês … parrêsias*) in the [doctrine of] two columns [of opposites] to be found in the Pythagoreans.[341] However, [Aristotle's] point (1003b22-5) that unity and being are the same and together are one nature, as are principle and cause, the part-less and the smallest, and the other matters for which [our] apprehensions (*epibolai*) and accounts differ,[342] but what underlies is one: this point presents many difficulties, both for those who can see things [as they are], and for those who wish to preserve consistency (*homologia*)[343] in what Aristotle says.

For the former may consider that that which is one is unified by deriving its unification from elsewhere,[344] that this is clearly from the One beyond being,[345] which itself is not in need of being, but being purely one, gives unity to all [things]. Without [this unity] things can neither be constituted at first, nor, once constituted, can they persist, so that the goodness of the One prior to all things sets up for them this power which sustains and protects them.

As for the latter, why, they will say, are unity and being the same? Is it for the reason that if, on the one hand, 'one' and 'man' do not always go together, but when there is 'man', there is 'one', yet when there is 'one', there is not in every case 'man', or 'sun', or 'world', or 'intellect', on the other hand, however, where there is being, there is one, and where there is one, there is being?

But first of all, this [last point] has not been conceded. For unity is both above[346] being, and with being, and below being, as in the case of matter and privation. And in general[347] why must it be that if they go together, they do not differ from each other in terms of priority and posteriority? Then, how is it that <if>[348] substance is <unity>, then number will not also be substance, and this in the case [of number which is] in what is numbered? For if Socrates is both one and essentially being, with Plato they are both two and essentially beings, and so on. Therefore, as substances, in becoming many, do not change into accidents, but will again be substances, thus also the unity that is constituted in each of these will produce a substantial number. For if the number of ten men is an accident, the ten substances are not accidents, but are substances of themselves. So for one, being one will be an accident, but it will be itself substantial. Even then in the case of Socrates, being would not be the same as unity, if it is the case that the former, as something composed and multiplied, is simply being and substance, whereas the latter as progressing appears as a non-substantial (*anousios*) number.

Again, from another perspective (*epibolê*), if in relation to the substrate unity is the same as being, the opposite to being, non-being, will also be identical with the opposite to unity. So multiplicity will be non-being. But the many are beings, so beings will be non-being! But if someone, basing himself on the endings of words, comes up and says that beings are non-being, since they are not 'being', but 'beings', he will be attempting to deceive us (*sophizesthai*). For it is not the case that because multiplicity is a multiple of what is one, that it is said not to be, but because it is a privation of being.

But perhaps in response to this someone might say, on Aristotle's behalf, that neither is multiplicity complete privation of unity, but proceeds from it, whereas if you were to deprive multiplicity of unity, completely and in every respect, in fleeing, being goes completely towards non-being and the abyss of what absolutely and in every sense is not.

For my part, I would like them to say and think these things, for these things are somehow more Pythagorean and Platonic and what is most cherished by Parmenides, and at the same time leading somehow to the conception that all things exist and are preserved by the One.[349] But I inquire of them further, whether they think the one man is more worthy than the ten men. If the one man, then how, multiplied by ten, does the value of the thing not become greater? But if it is the ten men, how can the one be form and being, and the many privation and non-being? But if they would say that the ten men, in relation to number, are deficient in regard to the one man, but exceed in relation to what is substantial in man, then in the one man his being and unity would not be the same. For otherwise, in being multiplied, one part would not become more worthy, the other of lesser value.

Again, if multiplicity does not stand as privation in relation to unity, but they are opposites in the category of relation, following indeed what Aristotle himself allowed in book 10 (6, 1056b32-34) of this treatise, it would be necessary also that these men be opposite to the one man in this way. But nothing is composed of opposites to itself, thus neither are the men opposite to the man, nor is a unity of this kind, out of which is composed the multiplicity, opposite to the multiplicity. Again: man as man and being as being do not differ from men and from beings, but unity as unity differs from multiplicity as multiplicity, so it is not the same thing for man to be one and to be. So unity and being are different, even if unity exists with being, since nothing can be without unity.

But let these difficulties be posed as resulting from the teaching of this demonic[350] man [Aristotle], a teaching which is sometimes discordant (*diaphônon*) since it both arouses to the higher and more inspired (*epoptikôteran*) philosophy, and [yet] retains what is more approved by the majority due to the force of habit. For having begun now to sing the praises of unity in a Pythagorean and Platonic way, saying it is substantial and giving it the leading role in the better of the two columns,[351] yet because he failed to master its nature as transcending all beings, he found himself in opposition to himself. For it belongs only to what is above beings to transcend multiplicity. Since, if multiplicity is not lacking in another respect with regard to the unity co-ordinate with beings, how would it be of less value than the One, if it is neither produced by it nor is each member of the multiplicity second to it? But let us go on to what follows.

61,1 So it is his opinion that being and unity are the same in substrate, even if the accounts of them be different. But if someone were also to give the same account for both, he denies (1003b24-5) that this will make a difference, for this concurs rather with his whole purpose. For it will be shown more fully that what comes under unity is all the more suitable [as a subject of inquiry] for the philosopher to the extent that he deals with being as being. What then are the proofs that unity and being are the same?

First (1003b26-7), that 'man' and 'one man' and 'being [a] man' show the same thing. Even if I say 'being one man', I do not say something different (for what is doubled, even if it be tripled, indicates the same thing). Second (1003b32-4), that the substance of each thing is not one *per accidens*, but in the way that it is precisely a being, so is it precisely a one. So that unity is, in as many ways, as being is.

The Essential Properties of Being
2. Unity, Multiplicity and their Species

Since both [unity and being][352] are the same, in examining substances and all beings, the philosopher will examine the differentiae of unity. The differentiae of unity, what are as it were its species (*eidê*), are: the same, the similar, the equal, the straight and, as a whole, the better of the columns,[353] just as the more deficient column follows multiplicity. Our philosopher himself dealt indeed specifically with this, making a selection of all the contraries[354] and arranging some under unity, others under multiplicity.

If therefore philosophy as a whole will be about substances and about the differentiae of unity and multiplicity, its parts will be about the kinds of substances.[355] Of substances one [kind] is intelligible and eternal, the other sensible and destructible. Therefore first philosophy will be that concerned with intelligible and unmoved substance, that part of philosophy which comes after this and which is second dealing with nature. For as being and unity have species (*eidê*), one primary, the others subsequent to it, thus also will the sciences be ranked corresponding to beings. For philosophy includes in itself the prior and posterior, like mathematics. Thus, as the arithmetician is prior to the geometer, and he to the astronomer, and he to the engineer (*mêkhanikos*), for example, so also are the species of philosophy as a whole arranged in the series: first, intermediate, last.

Since it belongs to one science to know what are in every way opposites to each other, in knowing unity it must know multiplicity. For multiplicity is the negation or privation of unity.[356] These oppositions differ from each other in that negation is true of all that which is besides unity, when unity is removed (for 'not horse' is true of all that which is besides horse).[357] But it is not so in the case of privation. For it is not simply because some state is absent that privation comes <to mind>[358] (for being blind is not the fact of not having sight, or else a stone would be blind), but it is because this state is absent from that which is of a nature to receive it. Thus it is not everything which does not hear that is deaf, but an ear, if it does not hear, since it is in its nature to hear. For one nature must underlie the state and the privation. And this difference between privation and negation is very great. However, whether multiplicity is a privation or a negation of unity (since it is possible to consider unity in terms of each of these), it belongs to the same philosopher to know unity and multiplicity. For in general it belongs to the one science to know things which are opposite to each other with respect to any kind of opposition.

But if the philosopher will know unity and multiplicity, he will know not only the same, the equal, the similar, as differentiae of unity, but also what is opposite to these, the different, the dissimilar, the unequal, which come under multiplicity. And not only these, but also whatever, through these, comes under unity and multiplicity. For the white and black, the sweet and bitter, and all things of this kind come under the primary opposites, because [i], as I believe, the sweet and white and all that which is more related to form come under unity, the bitter and black and all which is related to privation coming under multiplicity, or because [ii], as Alexander explains,[359] that which is separative comes under multiplicity, the opposite to these coming under unity. And contrariety itself we will relate to multiplicity, love (*philia*) coming under unity. 'For contrariety', Aristotle says, 'is a difference, and difference is otherness'(1004a21-2).

Now, my good man, you say all this very well! Why therefore have we[360] not also included the One and the Indefinite Dyad among the principles? Why did we fight with the Love and Strife of Empedocles? For we now introduce the same things and the same powers among the principles, under different names, not even changing the name of the One ['unity'], but renaming the Indefinite Dyad 'multiplicity'. All this is indeed very well said. But since it is necessary that the principles transcend that of which they are principles, you should take unity and multiplicity as transcending that which share in unity and multiplicity here: it is in this way that we will consider things in a way that is more worthy and not disagree (*diaphônêsomen*) with ourselves.[361] This will come up again, if appropriate.

Having shown us, up to this point, that it belongs to one science to consider all beings and that which comes under unity and multiplicity, [Aristotle] resolves an objection which questions how, if they are said in many ways, unity and being come under one science. Aristotle says that since they are said by reference to one thing, just as what is 'medical' belongs to one science, so too do they. For all beings are related to substance and to unity in a primary sense. And so also if the same has differentiae, they will all be ordered to what is the same in a primary sense, that is, that which is the same in [the category of] substance. And so also [in the cases of] the different, the similar and the dissimilar. Therefore, having distinguished the many ways in which each of these is said, be it in ten ways (since they exist in all being) or in a different number of ways, we should co-ordinate what is in each category with the same, the similar, the different and the dissimilar in [the category of] substance. For some of them [are said]
63,1 in the sense that it, as the first, has them; some of them in the sense that it, as the first, first makes them (for substance has and makes the [things] which are said to be); some of them, [as related to it] in other ways. For either they have a relation to substance, or are an affection of substance, or something like this.

Thus is also resolved the difficulty[362] as to who it is who is to deal with the same, different, equal and unequal, and in general all those things which are essential properties in substances. For it is clear from what has been said that it is for the philosopher[363] to deal with these. And indeed, if it is not the philosopher, who will demonstrate either that Socrates and the philosophising <Socrates>[364] are the same, or different, or in one way the same, in another way different; or if it is possible for many things to be contrary to one, or impossible, whereas one [thing] is necessarily [contrary] to one? Therefore, since these things are the affections, neither of numbers, nor of figures, but of being as being, it is clear that the study (*skholê*) of these things belongs to first philosophy.

And it is not because they deal with these things that one might criticise the sophists, but because they do it with great ignorance, knowing nothing about substance. But one ought to have known that just as there are affections proper to number, the odd and even, and, in the relations between numbers, the multiple, the submultiple, and in general the ten relations of excess and deficiency;[365] and that just as there are [affections], both in a solid which is conceived by geometers and is unmoved, and in a natural and mobile solid (whether weightless and celestial, or enmattered and having weight or lightness); so also indeed are there certain essential accidents in being as being, about which nothing sound can be said by him who has not examined the matter. It is therefore only the philosopher who will treat of these matters in a sufficient way. But the dialecticians and the sophists, usurping this role,[366] pretend that it is they who discuss all beings. However, they are deficient in respect both to the philosopher and to the truth: the former,[367] due to their not having his science, but using an approach based on opinion (*endoxôs*) concerning that which is demonstrated by the philosopher; the latter, due also to their evil purpose. For the [sophists] seek almost the opposite to that sought by the philosophers: the latter aim at saving those who come to their lectures (*akroaseis*), whereas the former aim at deception and making money from those they have perverted. But dialectics has a better purpose than has sophistry and has a knowledge which is philosophical, even if the dialectician puts those assembled to the test (*peirastikos*), not teaching them (*didaskalikos*) as does the philosopher. But the sophist is the image of an image, for he imitates this skill of testing (*peirastikê*), but is not able to do even this.

Having shown, therefore, from the cases of <sophistry>[368] and dialectics as imitating philosophy in dealing with all beings, that philosophy has knowledge of all being as being, Aristotle also makes use of another argument of the following kind (1004b31). All natural philosophers and those who seek a knowledge higher than theirs see the principles of being as contraries. He then adds the opinions of the

64,1 more ancient thinkers: the Pythagoreans with the odd and even, Parmenides with fire and earth, Plato with the limit and unlimited, Empedocles with Love and Strife. All these and all other contraries come under unity and multiplicity. But these are the object of one science. Thus the perfect philosopher[369] deals with all, whether unity and being be said univocally of all beings, or – and this is nearer the truth – they belong to what is said in many ways, those coming under unity and those coming under multiplicity being referred to one [thing] and one nature. And thus neither do we suppose that unity is separated from all things, nor do we say that it is predicated univocally[370] of things. And in general we bring under one science all of the attributes of unity and of being. For some [of these] are said by reference to one thing, others as being in a series, like numbers and figures. For it does not belong to the geometer to treat of the same, the similar and the perfect,[371] but if sometimes he names triangles as being 'similar', he takes these on the basis of a hypothesis and definition. For hypothesising[372] that similar triangles are those which having equal angles have the corresponding sides in proportion, he shows thus that these triangles are similar. But he is not able to say if such things should be called similar or not.

[Aristotle] then adds a conclusion (1005a13) to these arguments, in which he says that it belongs to the same science to consider both all being as being and its essential properties. And he adds (1005a16) that the same science will also know about the prior and posterior and about genus, species, the whole and part. For all of these are opposed to each other, and all opposites are included in unity and multiplicity, which, it has been shown, are matters especially for the philosopher.

Now we should again tell Aristotle that if that besides unity[373] indicates the other [member] of the [pair of] contraries, there must be another 'one' prior to all contrariety, in which all [things] participate, including those which belong to the other column, that which is the principle of all things. For the principles should start, not from a division, but from the One, since, as you say very well [Aristotle], 'the rule of many is not good' (12.10, 1076a4). And indeed the better Pythagoreans[374] do not start from opposites, since they place before the two elements[375] a transcendent One, cause of all, in relation to which nothing is opposed nor claims to be in any way of equal strength. Let this much suffice at present concerning these questions.

Chapter 3
The Principles of Demonstration

From this point Aristotle now goes on to investigating if the philosopher will discover the principles of demonstration, just as he does those of substance. And he states explicitly (1005a19-22) that it is appropriate for the philosopher to examine these matters. The arguments are as follows. The essential accidents of all being are demonstrated through these [principles of demonstration]. Thus he who shows these accidents will have more to say about these principles. For just as if we were to know only the accidents that belong to numbers through these [principles], the knowledge of these would belong to arithmetic, or if the accidents belong to magnitudes and shapes, [it would belong] to geometry, so, since the essential properties of all being are shown through these [principles], he who deals with being as being will treat of them. For indeed he alone will deal with them in a more universal way, since all other skills, to the extent that they deal with some being and not with all being, in distinguishing that which is fitting for them, reason on the basis of it, [e.g.] the geometer (that which are equal to the same magnitude are equal to each other)[376] and the arithmetician (if you take equal numbers from equal, those remaining will be equal).[377] Furthermore, going from the fact that no mathematician or other specialist attempts to speak of these principles, you will see that the study of these is suitable only for the philosopher.[378]

If however those who are keen students of nature often speak of these, they deceive themselves. For thinking that there are only corporeal things and that they deal with all being, they thus have something to say about the axioms. But we will remind them that they also are philosophers, as dealing with natural substance, but they are not first philosophers.

But if others, in attempting to speak of the truth of the axioms, offer demonstrations of them, they do this in ignorance of analytics.[379] For it is not for someone who wishes to know something to investigate the axioms, but to come possessing them already. Furthermore, if the other specialists know the most firm principles of the genus with which they deal, it is all the more necessary for the philosopher to know the most firm and fundamental (*arkhêgikôtata*) of all axioms. The most firm is that concerning which it is not possible to be mistaken; and the most fundamental is that into which all is analysed, as something anhypothetical. For a hypothesis is a principle in relation to something, but not a principle absolutely, being demonstrated through what is superior. But this principle must be the most firm, the most fundamental and clearest, so that none will be mistaken in respect to it. Such a principle is that relating to contradiction, for neither can the same [thing] be and not be in the same [thing] at the same time in the same respect and in the same way; nor can someone be disposed (*diakeimenos*) in this way, even if for the sake of argument this may happen [to him] in [advancing a] thesis (*thesei*);[380] nor can someone speak the truth in both affirming and denying [the same thing].[381]

Aristotle shows that it is not possible to be mistaken about this principle in the following way.[382] If someone thinks that the same is and is not in the same thing, he will have in himself contrary opinions, for opinions which correspond to a contradiction are contrary, as has been said at the end of the *De Interpretatione*.[383] Yet it is impossible that contraries be present at the same time in something. So it is impossible also for someone to be mistaken about this principle. And this is why all those who demonstrate something make use of this principle.

Chapter 4
The Principle of Non-Contradiction[384]
[The Principle cannot be demonstrated][385]

If however some of the Heracliteans or other natural philosophers believe that they also hold opinions of this sort and that things in the same respect may not be that which they precisely are, one must believe that they make such affirmations for the sake of argument. For we have clearly recalled that it is impossible for contrary opinions to be present in the same person. However it is not for this reason that we will attempt to demonstrate the truth of the axiom. For it is evidence of the greatest ignorance to think that *everything* can be demonstrated.[386] And indeed he who says this destroys demonstration, just as does he who says immediately (*autothen*) that *nothing* can be demonstrated. For each thing is destroyed in two ways, immediately or through the denial of principles. Thus, just as he who says that *all* things are generated destroys coming-to-be, since he does not grant that the principle of things generated is what is ungenerated,[387] so he who holds that *everything* can be demonstrated destroys demonstration, not allowing that there is something undemonstrated that is precisely that responsible also for the fact that what comes after it can be demonstrated.

But if in general there is something which cannot be demonstrated, what would it be, if it were prior to this axiom from which derives the power possessed by all other [axioms] which are not demonstrated? <For example,>[388] if equals are taken from equals, equals are left.[389] For what reason? Because if they were not equals, the addition of what has been taken away will make them all (*ta hola*) unequal. For the addition of the equal to the unequal will preserve the initial inequality. Therefore, the same things will be both equal and unequal, which is absurd. And I say, not that this is a demonstration, but that in bringing him who objects to the other axioms[390] to a more evident absurdity, we get him to reach contradiction. In general, if it is not agreed that they are equal, by the fact of supposing that they are not equal, what else do we obtain but the axiom of contradiction? And it is not we who suppose this, but he who objects. For what else would he expect us to suppose, when he does not agree that the same things are equal (one cannot even imagine this!) which are equal to the same thing? For if it were yet possible, let a not be equal to c and let both of these be equal to b. Therefore since a is not equal to c, and c is the same quantity as b, nor will a be equal to b. But it is equal. So the same will be equal and unequal, which is absurd. And in a similar fashion, regarding the other axioms, the refutation of those who make objections uses [the principle of] contradiction, since it too operates by reduction (*apagôgê*) to the impossible. And indeed all refutation brings back to contradiction, whereas he who supposes the opposite of the truth often uses contradiction and this in a most evident way. How then could this axiom be demonstrated, an axiom compared to which there is nothing more firm, better known and more fundamental?

[Refuting those who deny the Principle: the argument from signification]

Nevertheless, even if it is not possible to demonstrate this axiom, since whatever we might take so as to prove it would either be of equal strength to it or less well-known (*amutroteran*), it is still possible to refute him who objects to it, if only he agrees to use argument (*logos*). For let this man speak to us, if he has not ranked himself as an irrational animal or even as a non-perceptive plant.[391] Now [we ask him], is it not the case that each name signifies something?[392] If it signifies something, is it not absolutely necessary that it signify one thing, or many determinate things, or all things? But if it signifies all things, then it signifies nothing,[393] for that which does not signify this thing rather than anything else signifies nothing. Thus it remains that if it signifies something, it signifies either one or many things which are determinate. But if it signifies one or many things which are determinate, each of these [things] will have a proper name. Now if indeed one name signifies one thing, we immediately have what we seek. But if the one name signifies many determinate things, we will take among the many determinate 67,1
things those which are signified, and we will assign to each a proper name, for that which reasoning can separate, the one from the others, this it can signify by a name or account. So the one name signifies one thing. Since, even if it signifies many, this would not make contradiction. So that it would be in vain here to refer to things said equivocally.[394] If therefore, for example, one name, that of 'man', signifies one thing, man, it signifies nothing other than what man is. For if it were to signify also one of his accidents, such as white or musical,[395] it would no longer signify one thing. But in the same way as one might give the one name to man and horse, thus, if with the same name you will speak of man and white, this is no longer to give one name[396] to one thing. Thus, when I call this thing 'man', I show it to be nothing other than an animal which is land-going and biped.

If then I say the truth in saying that this is 'man', it is necessary that it be an animal which is land-going and biped. But if it is necessary that it be an animal which is land-going and biped, it is impossible that it not be an animal which is land-going and biped. But if it impossible that it not be this, then he who says that it is not will be speaking falsehood. So if it is true to say that he is a man, it is false to say he is not a man. Consequently, that concerning which an affirmation is true, the negation in relation to it is false.

One might well be amazed, in these things which Aristotle says, both by the focus (*epibolê*) of the argument and by the whole way in which the discussion is handled, being done with great wit and skill. But one might also note that if he who objects to the negation[397] cannot gainsay the other things which are said once he accepts discussion [with us], he might contest this point,[398] even if it is very clear. For, he might say, if it is necessary that this be an animal that is land-going and biped, it is not impossible for it immediately not to be [this]. For this is what he constantly says, that each thing is this and the opposite. Well, we will say that he does away with something which has been established, for the one [name] no longer signifies one thing. But he might say: this signifies one thing, not however the one thing as a whole but a part of it; what excludes the opposite being said of the rest? And this is so, [he says,] if the nature of things is stable and determinate. But if each thing flows and always changes, as the Heracliteans say, things will not be more in this way than not in this way.

In reply to him[399] we might say [i] that we will now set aside the Heracliteans, if indeed it is in this way that they reasoned. For, in paying attention to what they say, they will deny that names signify something. For as you cannot compress with your hands the blowing air, thus it is not possible to use names to show the unstable changing nature of things. It is for this reason, according to some,[400] that Cratylos, later, abstained from using speech, simple nodding in agreement or disagreement. However, if in general [ii] there is a determinate existence for things, there is also for each and every one, in every case, one conception (*ennoia*) and one name. For if there were not, each would require many names, which, however, would not show both the state (*hexis*) and [its] privation. For this indeed is really impossible. For as it is the case that if a state is present it is no longer possible for its privation to be seen in a substrate, thus it is not possible to reveal the same thing through the name of a privation and of a state, in the same regard. If this is so, neither would one express (*hermêneuein*) the same thing through affirmation and negation.

Therefore, taking up the argument again,[401] we will arrange it in the following way. If a name signifies something, it signifies some one thing. For if it signifies more things which are determinate, we will take what reveals each one thing.[402] But if the name signifies one thing, it signifies its being. But with its being does not come, I take it, its privation (for it is the absence of being that originates privation). Consequently, this may be expressed by an affirmation, this being impossible by the opposing negation, since it is impossible also for the state and privation of the same to be present at the same time in the same thing. For if it were true to say of this thing that it is an animal and is not an animal, this thing would have the form of an animal and the privation of the same animal. But it is not possible to have the privation if the form is not destroyed. Thus it will not be affirmed and denied in the same respect. For if it is not, in the same respect, man and white, but in one respect and in another, how could it be man and not man in the same respect? It would be as if one were to say that the form and the privation of the form are one. But if it is in one respect man, and in another respect not man, this is not a contradiction, since Socrates too is a substance, as man, and not a substance, as father or snub-nosed.

Has Aristotle then provided a demonstration of the point that a contradiction, in both its parts,[403] is not true? Now earlier[404] we have testified against this, that it is impossible to demonstrate this. But Aristotle attempts to demonstrate by refutation,[405] taking the other things that are conceded and taking this as true, that if it necessarily is the case, it is impossible that it not be, a principle taken as not sufficient to respond to him who says that a contradiction, in its two parts, is true. And we have tried to show that one cannot object to this, lest one reduce both form and the privation of form to the same thing. Now we have taken as true this point, that form cannot be together with its privation. But if so, then neither would man, as man, be not man. However, this principle is not as evident as the axiom of contradiction. Thus Theophrastus is right in saying that these people are most lacking in sense and violent who impose the necessity to give an account of this axiom.[406] For he who objects to it immediately abolishes the use of argument and demonstration. For if it is not more the case that it is this than its opposite, what need is there of argument and proof and deriving conclusions? Where is the fruit of this whole treatise, if he who tries to support the axiom is forced to have recourse to something that is lesser than it in evidence (*enargeia*) so as to prove it?

Having shown therefore that if an affirmation is true, its negation cannot be true, Aristotle says that similarly it will be demonstrated that
if the negation is true, the affirmation can never be true.[407] For if it is 69,1
true to say that this is not an animal, it is impossible that this be an animal; but if it is impossible that this be an animal, it is necessary that this not be an animal; but if it is necessary that this not be an animal, then he speaks falsehood who says that it is an animal. Consequently, if the negation is true, the affirmation will be false.

Having thus conducted the demonstration, it is clear that Aristotle, being able to go also through the two other hypotheses in a similar way (for if the affirmation is false, the negation is true, <and if the negation speaks falsehood, the affirmation will be true>),[408] saw that sometimes he who through malevolence (*panourgia*) objects to the axiom of contradiction will not agree that either of the statements is separately (*aphôrismenôs*) true or false, so that the other part of the contradiction may not be shown to be opposed. But in being asked 'is this an animal or not?', [the objector] answers that it is an animal and not an animal. And to 'is this a man or not?', he will say 'a man and not a man'.

So then we will tell him, Aristotle says,[409] first that you anticipate what we are looking for,[410] then that you bring into confusion what is proper to dialectic. For when I offered you one thing, you would not answer with regard to it, but add something else. For if indeed one might agree that the man is also not a man (for it is possible for him not to be a man in regard to one of his accidents), the question nevertheless concerned one thing. Thus just as if I were to ask you if Socrates is a man, and you were to say 'a man and snub-nosed and Athenian' and I said to you 'one thing is enough, the first, but if you want to enumerate his accidents, it is the moment (*hôra*) for you to go through all and in doing this you are doing everything rather than carrying out a dialectical discussion': so it is the opportune moment for you to hear, when you add 'not man', that the answer has become more superfluous than dialectical.

[The refutation continued: the predication of essential properties]

In addition to these arguments, Aristotle also uses against these people this sort of an argument (1007a20). Those who say that the same thing is and is not seem to eliminate, from among predicated terms, substance and what a thing is. For if it were possible in general that a thing be and not be the same, this might happen sometimes in the case of accidents, but it is impossible in the case of substances. For that which may or may not be in something is an accident of it, although this does not indeed happen in ways that are opposed. For neither when it is in it, nor how it is in it, precisely then and in the same way is it not in it. Nevertheless, since [an accident] sometimes is and sometimes is not [in it], contradiction has more room for itself in the predication of accidents than in the predication of properties which are essential and with regard to substance. But if these people eliminate substance from what is predicated, they also abolish the immediate premise, what Aristotle calls here and in the *Demonstratives*[411] the 'universal' premise. But in abolishing substance and the immediate premise, they make the predications

accidents, and these to infinity, which is impossible. For accidents are predicated in two ways: either as the accident of substance, as when we say 'Socrates is musical', or as the accident of an accident, 70,1
as when we say 'the white is musical', because both are accidents [present] in the same child.[412] Now let us set aside this last [type], since it is not [our] purpose to examine it. As for the first [type], it will not have more than two accidents in the <same>[413] predication (*katêgoria*).[414] For if I say 'Socrates is a doctor', I predicate one thing of one. But if I wish to expand this, I make the proposition 'Socrates is an excellent doctor'. But it is not possible to extend this further, for the proposition will no longer stay one. Thus, if a contradiction is true in both its parts, there is no immediate premise (since in general nothing essential will be in another), and if there is no immediate premise, the predication of accidents will go to infinity. For if someone were to say 'Socrates [is] a doctor', and I ask 'why?', he will not tell me of one of the essential properties of a doctor (for there are no such things), but says, for example, 'because he is white', and then I ask 'so why is he this?': going to infinity, there will necessarily be no end to the predication of accidents. However, this is impossible, for it is not possible to predicate more than two [accidents] of substance.[415] So never will a contradiction be true in both parts.

But, someone might say, what then prevents us predicating, if not more than two accidents of substance in one proposition, many and indeed an infinite number of them in the whole demonstration? However every demonstration is a definition, when the middle term is transposed, as Aristotle shows in many places,[416] and the definition is an immediate premise or in general a premise including what is defined.[417] Thus again one would be predicating more than two accidents of substance in one premise. This much on this matter.

[A conception of matter may inspire the denial of the Principle]

Having shown that those who say that a contradiction is true in both parts try to rule out both essential properties and that predicated with regard to substance, seeing[418] only predications with regard to accidents, Aristotle then shows (1007b26) that they themselves seem to be imagining (*periphantazesthai*) matter, but not at all that which is formed and receives definition and stability from nature, affirming very cleverly and with philosophical depth that having previously turned away from substances, their unstable and unreliable opinion now rejects forms. And indeed matter would seem to be more receptive of contradiction than anything else of what is, for matter is potentially all things and in actuality none. Thus it is man and not man, horse and not horse, each and not each.

But yet, in the truest sense that is, even matter does not accept simultaneous contradiction, for in one respect it is man, as being of a nature [to be man], and in another respect not man, in that privation occurs in it prior to the presence of form. But nevertheless contradiction finds room for itself more in matter than in other things, just as it does in what is accidentally predicated more than in properties with regard to substance.

How then are these people led to this opinion? If they were to say, like Protagoras, 'all that which might appear to someone, this it also 71,1 is',[419] they would immediately say that all things are in all and Anaxagoras' doctrine would prevail with them: 'all things together'.[420] For if the same thing is white and sallow and black, since it appears differently to different people, and admits of negations of these predicates, the same thing will be all things: this is the matter (*hulê*) of beings.

But if they were to say that that [thing] of which the one part of the contradiction is stated, of this is stated the other part, which does not relate to all things, they are also forced in this way to assign the contradiction as a whole to each [thing]. For if man is not man, much more so is he not a trireme, not a wall, not iron, for the negations of other things suit him even more than does the negation of himself. Thus if he admits the negation of himself, he also admits the negation of the others. But if, of that of which affirmations are made, of this also negations are said, then the man will be both all things, since all things will be truly affirmed of him, and nothing, since the negations of all things will fit him. Thus again will Anaxagoras' doctrine prevail, 'all things together', he speaking in this way on account of [his] image (*phantasia*) of matter.

[The refutation continued: the Principle of Excluded Middle]

Aristotle furthermore says (1008a3-4) that these men are also forced to resist the other axiom of contradiction.[421] For if the negation is of that of which the affirmation is made, it is necessary for them to deny also that the [one or] other part of the contradiction can be said of anything. For if the same [person] is man and is not man, the negations of these [predications] would either be two or one: the negations of these as one, [as in] 'he is neither a man nor not a man'; or as two, [as in] 'he is not a man, he is not not man'. Whatever prevails, 'man' is not predicated of him, since he is not man, nor is 'not man', since he is not not man.

So the argument here has undergone an affection almost like Protagoreanism. For Protagoras, in maintaining that every opinion is firm and true, is nonetheless forced into saying that every opinion is false, because of those people who contradict each other. But they all, in attempting to unsettle the determination and stability of things, have overturned[422] themselves.

Yet Aristotle will not even give them this. Rather, he tries to force them either into allowing some determination in things and not condemning things to disorder and indefiniteness, or else not holding themselves to be rational or in general perceptive beings. For he asks them (1008a7-11) if each being accepts all contradictions, or not (some accepting some contradictions, others accepting others). For if each thing does not accept all contradictions, then a determination would already appear in things. For this fact, that these things are in these, but those are not, belongs to some discrete property. And if not all affirmations, but all negations apply in respect to the same thing, then something firm would be found as *not* being in them.[423] Yet it is ridiculous to give firmness to privation, but not to form. But this happens if they say that all affirmation comes with its negation, but not all negation with its affirmation. For how would it not be absurd for the negation to be more firm, more unadulterated and more knowable than the affirmation? But yet they allow something even to be determined,[424] as either a privation or whatever.

[In what ways are all affirmations and negations true?]

But if they say, in respect to the same thing, that all affirmations and negations are true, they should be asked[425] whether [i] each, when it is said, is also true by itself (for example, 'this is a man' and again 'this is not a man'). Or [ii] is neither of these statements in itself true, but only when both are taken together do they make truth ('it is a man, it is not a man')? Whichever of these [options] is chosen, it will lead to absurdity, whether [ii] the positions (*theseôn*)[426] are taken together or [i] each by itself. For if [ii] the statements are not true, taken by themselves, first they signify nothing, for an affirmation which is neither true nor false signifies nothing. Then there will not be something among things, if indeed it is not possible to affirm or deny something about one of them by itself. But if there is not something, then there is no man, if, that is, man is something among beings. Thus will there be neither what will speak nor what walks, if man is not.[427] But even more paradoxical than this again is that each one thing is all things. For if the same thing is man and not man, wall and not wall, it would be all things at the same time, and the one thing would be nothing and all things, if it is the case that by themselves affirmations and negations signify nothing, but are true only if joined together. But if [i] they were to separate these from each other and say that they are true, the same thing would be all things, by reason of the affirmations, and nothing, by reason of the negations. In consequence, this is what is common to the two hypotheses (*hupotheseôn*), whereas what is particular to the one hypothesis is to say that affirmations and negations which are taken separately signify nothing, while what is particular to the other is to say that all are true and all false. For if all affirmation is true and all negation is true, he who says anything speaks the truth in what he says, but speaks falsehood because the opposite to what he says is also true, so that he also agrees that he belies himself. For were he to say that this is a man, and agrees that this also is not man, he thus agrees that he belied himself when saying as a separate [statement] (*aphôrismenôs*) that what is not a man is man. And Aristotle says that it is clear that such a person is unfitted for sharing in discussion. For he immediately contradicts himself in abolishing what he poses, and again even destroys the contradictory statement itself, so that nothing definite might remain, thus driving himself into the indefinite, the obscure, the unordered and unknowable.

Nevertheless he asks them again:[428] does he who thinks that this substrate is disposed in this way, such as to be heated, say what is false, just as does he who denies this, whereas he who thinks and says both speaks the truth? For if the latter does not say the truth either, how can they still say that the nature of things is like this, 73,1 such that they are and are not like this? But if neither of these speaks the truth – lest what is determined slip into things with the truth – is it however rather he who speaks both parts, or he who speaks one part of the contradiction? For if it is the latter, something determinate would already appear among things, <of which>[429] he who comes nearer to it will be said to say more the truth, whereas he who misses it will be said rather to be completely wrong. But if they dare to say that all statements speak the truth in the same way as they speak falsehood, well then they abolish discussion, teaching and demonstration. And if they were also to suppose things in this way, they would appear in the end to be neither men nor in general perceptive [persons], but rather as living the life of plants.[430]

[Impossibility to act, if Principle is denied]

But perhaps it is impossible to be disposed (*diakeisthai*)[431] in this way, as impulses and actions show,[432] since these people run to Megara if they need the advice of the Megarians, they are cautious of a well, not thinking also that it is not a well. And in general they avoid ravines, precipices, all that is dangerous, not thinking that all these are also salutary, just as they again search out that which is salutary, but do not avoid it as also being the cause of destruction. Thus to all it is clear from these things that neither those who later were called the 'Ephectics',[433] nor those before Aristotle who supplied these arguments withheld their opinion on everything or opined opposite things, but they held to certain things without quibble and unhesitatingly, in particular about what is better and worse, not perhaps on the basis of knowledge, but on that of correct opinion. Thus[434] it would have been more helpful if they had cared for themselves (*heautôn epimelêthênai*) as not yet reaching perfection as regards truth. For as the sick person ought to be more attentive to matters of health than the healthy person, if he wishes ever to find an end to his unnatural condition, thus it is right that he who has not yet acquired knowledge should exert himself with a view to acquiring it so that he may at some time become able to say the truth.

[Are all statements false to the same degree?]

Aristotle also argues in this way against those who say all things are false (1008b31ff.). Let it be, he says, that all things are false. Are they all false in the same way? He who says that 2 is even, does he speak falsehood in the same way as he who says that 3 is even? Or does he who says that 4 is 5 go wrong in a way comparable to the way in which he who says that 4 is 1000 does? Or is it not rather the case that, in the first example, one speaks truly, the other falsehood, whereas, in the second example, one errs a little and the other a lot? But nevertheless let all speak falsehood, following the hypothesis: would one not say that one has less missed what is the truth itself, the other more? And if indeed this were agreed, then he has already said something as being true, in relation to which one of them has erred more than the other. But if someone does not accept the truth as existing when we argue in this way, we might at least win this from the argument: that it is not the case that all things are related in the same way to truth or falsehood (from which it would have followed that nothing is determined). Thus freed from the purest madness, we have forced them to accept an obscure determination (*horos*) in things, which, however, is something useful achieved through our questioning.

This is what Aristotle says in his attempt to show, through what is more true, that there is truth itself. I, for one, admire the man for his sharpness and demonstrative science, but I would have wished that the philosopher give voice to the same words concerning the circle, the equal, and suchlike.[435] For if, in the heavens, there is the circle and what is more equal, then there would also exist the circle itself and the equal in itself. But let us defer these matters to another lecture (*skholê*).

Chapter 5
[The relation with Protagorean relativism]

Aristotle again identifies Protagoras' opinion with those who say that [both] parts of a contradiction are true together. But one should know that if perhaps Protagoras' opinion and all [such] aberration follow from what these people say, since they produce confusion in everything, the error concerning contradiction does not follow from what Protagoras [says]. For it is clear that if he said that, for the *same* person, the same thing is sweet and not sweet, then he would fall clearly into the error concerning contradiction. But if he says that for *this* person the same thing is not sweet, but for *another* person it is sweet, and that in general truth and all things are relative, then he would not throw the axiom of contradiction into confusion. Let there be two persons who seem in contradiction, the one saying and thinking that this is white, the other that it is not white. If he were with them, Protagoras will say that these [claims] are clearly not in conflict with each other on this point, for each person says the truth, the object being white for the one, and not white for the other, this not being a contradiction, no more than if someone were to say 8 is double and not double, the double of 4 but not the double of other numbers. Let this suffice as to this difficulty.

[Perceptibles may mislead those who deny the Principle]

Aristotle says (1009a17-22) that there are two kinds of people who oppose the axiom of contradiction. There are those who are in need of more disputatious refutations, those who champion this position for the sake of argument and impressing the crowd. To these he has already responded adequately. Then there are those who are completely confused by certain puzzles and are led astray as it were by confusion in thinking. These he will consider as also deserving arguments of a more explanatory kind, freeing them of their discursive bonds and calling them back to notions (*ennoiai*) which [they possess] by nature.[436] Neither is the doctor who does not know the cause of a sickness up to curing the patient, nor is he who cares for souls, who has not first examined[437] the soul and investigated the causes of its false beliefs, able to bring us to perfection. It is for this reason that the philosopher, too, in investigating the causes for these people having false beliefs on this point, says that they are led astray by the alterations of perceptibles (1009a23), their continual changing and turning into the contrary. For seeing that from the same things are generated the hot and cold, wet and dry, hard and soft and all contrarieties, and taking it that nothing comes from what is not, [these people] supposed that the underlying nature was both hot and not hot, both wet and not wet.[438] [This] persuaded [them][439] to say that in all homoeomerous reality all things are present in actuality. One should posit that contraries are not simultaneously in actuality, but in potentiality. Yet they posited them as in actuality, something which cannot sensibly even be conceived. So, 'being' being said in two ways, in actuality and in potentiality,[440] they supposed that 'being' is said in only one way. And then again, 'non-being' being said in two ways (for non-being is both what absolutely in every way is not, and what is not yet, which comes to the same as what is in potentiality), they supposed that it also is said only in the former way.

To them we will therefore say that what becomes, comes to be both from being and from non-being, from being in potentiality, from non-being as not yet being. For man comes from what is potentially man and from what is not man: for such is the seed and, far prior to it, matter. Therefore the same thing is man and not man: not those already in actuality, but what are potentially. So much then as to what concerns what comes to be.

But one must think also,[441] among beings, of the most divine, the intelligible and unmoved, from which all that is potentially is excluded. For there are pure actualities there, in relation to which there is not even a hint that the same thing both is and is not. For all are limits and determinations and the indeterminate has been driven from there.

Now having shown that those who think that a contradiction is true in both its parts are led astray by the things that are in coming-to-be, through not separating what is potentially from what is in actuality, the philosopher says that the Protagoreans also derive from perceptibles, which appear differently in different places, the supposition (*hupolêpsis*) that what appears to each person is what it is. For one ought not to judge things according to the number, great or small, of those who make statements, nor would any other way of discriminating between them be conceivable.[442] For clearly the same things appear to us to be sweet, if we are healthy, and bitter, if we are sick. And the same things appear to us to be bitter and to other animals to be sweet, such as the olive-branch. Hence Democritus, too, declared that nothing is true, <or>[443] it is unknown to us,[444] and Empedocles and Parmenides identified what appears with the truth.[445] [Aristotle] reminds us also of the argument of Anaxagoras,[446] which claims that things are such for us as we ourselves might happen to be. And Homer says that when Hector went astray in his wits, he nevertheless[447] had intelligence, if of another kind, not that of someone who is healthy: 'for he lay thinking other thoughts'.[448] Now it is clear that the ancients did not say these things in the way Aristotle takes it. However he says that, since this is the disposition (*diakeimenôn*) of those who seem to be leaders in the philosophy of the Greeks, it is likely that those who enter the hunt for truth will be discouraged, lest perhaps they be pursuing the unattainable.[449]

Aristotle responds therefore to all of these people as follows (1010a2ff.). First: beings are not just perceptibles. For in these there is much indetermination deriving from matter and what is potential 76,1
and the appearance (*phantasia*) of opposites, so much so that Cratylos, too, seems to have a point in finding fault with his teacher for

saying that we cannot fall twice in the same river, since [he says that] we cannot do it even once.[450] Then (1010a23ff.): even in perceptibles themselves, there is something that changes and something that is stable, for what changes is quantity and the other affections, whereas Socrates' form remains,[451] from birth to death. And yet even what changes in him is not without stability in any respect, for it has not finished changing, but it [continues to] change. But it is clear that the form persists, so that it has no more room for 'Socrates' and 'not Socrates'.

Now all this concerns enmattered things. But it would have been right to look up also to the heavens, which are full of divine impassible substances, which receive nothing indeterminate from coming-to-be, nor leave those things here below entirely bereft of limit and coherence (*sunokhê*). It would also have been right to reach a conception (*ennoia*) of the intelligible and unmoved causes, from all of which it would have been possible to deduce that not everything both is and is not in such a state, but that the intelligibles and celestials are stable, always maintaining one determination and one order, whereas those things which come to be, being small both in quantity and value in relation to the whole heavens, never accept contraries in actuality, but in potentiality on account of matter. For if opposites were in actuality in generated things, no movement would be observed in them, nor any kind of change. For what is now white both has come from what is not white and changes into what is not white. But if it is in the same respect both white and not white, it is both ungenerated and indestructible and in all respects unchangeable.[452] For it does not have that to which it will change. Thus it will happen that those who say that opposites are in actuality in generated things will bring all things to a halt. But if change is abolished, there will be no generation and nothing would be generated, for all will be in all.[453]

[Refutation of Protagoras]

[Aristotle] also refutes Protagoras, who claims that all that appears is true,[454] since he has recourse to two false premises, 'all appearance is sense-perception', 'all sense-perception is true'. The conclusion is clear, but the minor premise[455] is wholly false and the major premise is false in one respect. For each perceptive faculty is more in control of its proper objects when what is intermediate does not prevent this. Then again he says (1010b3ff.) that those people are ridiculous who raise difficulties as to those who are nearer to perceptibles and those who are healthy and vigorous and awake, whether they judge better the same things, or if it is those who are at a distance, who are sick and weak and asleep. For *they* are not disposed (*diakeintai*) in this way (their actions show this).[456] And they force us to demonstrate the principles of knowledge which we have received from nature, [namely] sense-perceptions and common conceptions (*koinai ennoiai*),[457] principles in relation to which we are disposed (*diakeimetha*) in a way that is superior to all demonstration. Therefore the opinion of the specialist concerning things both present and in the future will be more authoritative than that of ordinary persons, even if there be many of them.[458] And the judgement of sight concerning colours is superior to that of taste, just as taste is superior to sight when it comes to flavours, and so also for each: as to its proper faculty of judgement (*kritêrion*), it never says that what it judges is both such and not such.[459]

But what then? Is not the same wine now stated by taste to be sweet, and now bitter? But first this happens at different times (this is not a contradiction). Then, taste does not judge essentially the wine, but sweetness, which no longer occurs if the wine goes bad or if some flavour combines with it on the tongue and destroys it, since as long as sweetness remains and affects the tongue, taste expresses the same judgement about it.

In general, however, the arguments which introduce these theses (*theseis*) attempt to abolish both substances from things and necessary premises from demonstrations. For there is no room for 'such and not such' in properties with regard to substance or in necessary premises.

[Aristotle] shows, in addition,[460] that those who say that all appearance is true generate perceptibles in conjunction with sense-perception. Therefore if animals and perceptive faculties are destroyed, there will not be perceptibles either, so that all perceptibles[461] will not exist according to these people. And yet what moves[462] ought to have been prior to what is moved,[463] but in this case perceptibles are generated and destroyed in conjunction with the perceptive faculties.

Chapter 6

But if, overcome by these arguments, they ask (1011a5), who will judge who is healthy or not, who is a person of intelligence (*phronimos*) or not, and if they are the sort that are having difficulties,[464] we will say to them that for some things we are naturally judges, as tasters of what is sweet, for other things we will leave the judgement to those who know, whom we recognise as being such from their deeds. And we also dispose of common conceptions (*koinai ennoiai*) which allow us to comprehend many things. And one must not make the principles of knowledge things to be demonstrated.

But if these people are of the more disputatious sort,[465] we will also be advocates of the truth with them. For we will say that they seek from us the impossible. For that which we wished them to do, turning them in the direction of contradiction, this they have accepted as presupposed.[466] Thus, just as in the case of those who throw themselves from a precipice or kill themselves in some other fashion can no longer be corrected by punishment, in the same way it is not possible to turn around[467] those who, from the start, say the most absurd and senseless things. For avoiding that which a refutation wished to offer them by way of therapy, they have accepted[468] to become incurable, and they have also tasted of this joy of the wisdom so much desired:[469] to have become more powerful than those who attempt to cure them.

Yet it is possible[470] to use the following arguments against them. If all that appears is true, everything is relative, so that this would be true at this moment through this perceptive faculty for him acting in this way. If not, nothing will be found to be more true than false. If then all appearances are true, all appearances are relative. And if not all are relative, but there is something which both is essentially true and has a determinate nature, then not all that appears is true. For the same thing appears as white to this person and as not white to another. Now if both of these persons speak the truth, then things are relative, for it is clear that the object is not simply white, but is white for this person and not white for another. But if it were to have some definite nature and were essentially white or not white, then it is clear that one of these persons is in error. Thus the conditional argument (*ho sunêmmenos*) is true: if all that appears is true, all is relative; but not all is relative, as I will show; so not all that appears is true.

Aristotle shows that not all is relative in the following way. If the objects of opinion (*doxasta*) are themselves present, corresponding to the object (*kata to hupokeimenon*), with the opinion, nothing unexpected would occur, nor would what is opined remain unaccomplished.[471] However each of these is clear: for both 'that which was opined has not been accomplished'; and 'a way was found for what is unexpected'.[472] So the object of opinion is not produced with opinion. If so, then not everything is relative. Furthermore: if all things are relative, man will not be both he who opines and he who is the object of opinion. For what is relative belongs to something *else*. Thus, if it is the same thing to be opined and to be man (for this is what the hypothesis holds: to be he who opines and to be man), it will not be said in relation to something else, a point which both completely overturns the hypothesis and conflicts with the truth. Again: whatever you take among things that are relative, you will find that it is said by reference to one thing, or at least in general to determinate things. But if what opines is said by reference to what is opined, it will be said by reference to an infinity of things.[473] For what is opined will be a horse, dog, deer and every [sort of] plant. If then the being of each thing were different, and being opined an affection common to them, what opines would be said by reference to one thing, because by reference to the affection common to animals and plants. But if what each thing is and its being opined are the same thing, if, that is, their form is produced in conjunction with the opinion, what opines would be said by reference to indeterminate things. So, not all things are relative, nor is all that appears true.

Chapter 7
The Principle of Excluded Middle

In saying all these things in response to these theses (*theseis*), the philosopher not only confirms that which concerns contradiction, but also discusses other axioms, such as that it is impossible for contraries to be present in the same thing in the same respect in the same way,[474] and that there is no intermediate[475] in a contradiction. For if there is something intermediate, either it will in every case be true, or false, for this is the case for every declarative statement. But it belongs only to affirmations and negations to be true or false. So there is not something [intermediate]. But how could [affirmation or negation], only, say the truth? Everyone who speaks the truth either says that what is is, or says that what is not is not.[476] This is [true] affirmation and negation. But how could they be false? Everyone who errs says that what is not is, or that what is is not.[477] The first is a false affirmation, the second a false negation. So there is nothing intermediate between an affirmation and a negation, neither in the sense that grey is said to be intermediate between white and black, nor in the sense that what is neither horse nor man is said to be intermediate between them.[478]

Chapter 8

[Aristotle] says that one should take a position in the same way against the other theses. For nothing is such as to be said just in one way in relation to all things which are, so not all are true and not all are false, whether it is Heraclitus and Anaxagoras who lead in the arguments, or others. For indeed such arguments refute themselves, in addition to other nonsense (*teratologia*).[479] For if all things are true, then that which they say that you say falsely, is completely true. And if all things are false, then even this doctrine (*dogma*) of yours is false. But if all speak truly (besides him who contradicts this wonderful argument) or all are in error (except for him), then there will be many who speak falsehood according to him who says all is true. For there is not only he who contradicts the statement, but also he who approves of him, and he who praises him in approving, and so on to infinity, so that many will be in error. And again according to him who says that all is false there will be many who say the truth:

himself, and those who join in with him, and those who praise the praisers, and no small choir (*khoros*) of truth-sayers will follow.

Nor will we let off someone if he says that all things are moved, or all things are at rest,[480] not even if someone says that now all things rest and now all things are moved. For we will say to him that it is not possible to make one statement (*logon*) truthfully for all things. Thus we will say that some of these things are unmoved, such as the intelligible causes, the poles and suchlike, some of them move eternally, such as those in the heavens, and some are sometimes moved and are sometimes at rest, such as things in coming-to-be.

Conclusion

Having discussed this much in this book, Aristotle brings to an end his teaching on these matters, having taught that which is the area of inquiry of the first philosopher, also defending the axioms concerning contradiction, furthermore consolidating as far as possible other principles both dialectical and natural, and in addition to all this, recommending this: to admit no single statement which applies one thing to all things, unless it is said equivocally. One should accept the statement 'all things desire the good',[481] for 'desire' is equivocal. One should[482] agree to the statement 'all beings are', for 'to be' is not the same thing for all things. But no univocal predicate is to be accepted as covering all things, since they are so multiform, both those in the cosmos and those beyond it.

Notes

1. *logikôs*: this adverb may be translated with 'verbal' here, since in Syrianus' view many of the difficulties raised by Aristotle only work on a purely verbal level: once attention is paid to the full range of what the words may signify, the difficulties disappear. See below 33,17; 45,34; 52,9-10 and Madigan's translation (with footnote) of Alexander *in Metaph.* 3, 177,9.

2. We have added this and the following titles of sections. The general title of the work (given above) is found in MS C.

3. On the goal of metaphysics, see the Introduction, above p. 3.

4. Syrianus distinguishes between the presentation of a series of problems in Aristotle's *Metaphysics* 3, chapter 1 (to which corresponds the first part of Syrianus' commentary) and Aristotle's elaboration of the problems in terms of opposing arguments in the following chapters (to which corresponds the second part of Syrianus' commentary) in an order not always the same as that of the series in the presentation in chapter 1. On the question of the differences between the order of problems presented by Aristotle in chapter 1 and the order he follows in the following chapters see below n. 78, for Syrianus' explanations, and Luna, 2004, 52-63 and 68-71 (who shows how Syrianus' treatment of the question influenced modern commentators).

5. MS C has *dia logôn*, corrected by Usener to *di' oligôn*.

6. i.e. in particular in *Metaph.* book 12.

7. We have inserted in our translation the division into chapters of Aristotle's text.

8. See above n. 4.

9. i.e. metaphysics.

10. Following Usener's correction *prôtôn* for *ontôn*.

11. i.e. in his knowledge; cf. *Metaph.* 1.2, 982a19-23.

12. cf. Aristotle, *Metaph.* 1.2, 982b4-5.

13. *pasi de kai pasais* (feminine); should one read 'all men and women'?

14. On these axioms, see below 18,9ff.

15. i.e. the metaphysician.

16. The sciences (for example, physics) which receive demonstrative principles or axioms from first philosophy do not know these principles or axioms by virtue of demonstration, but by virtue of 'belief' as having accepted them from a higher science. However, in this higher science, they are not known by demonstration, but by a higher form of knowledge ('above' demonstration), receiving them from transcendent Intellect; see below 6,5; 19,5-8; 22,30-1; 76,34-5.

17. MS C has *kai*; Jaeger's text of Aristotle has *kan*.

18. On the term *problêma* see Luna, 2004, 53-4.

19. i.e. the sciences using these four methods, on which see 4,26-9.

20. cf. Plato, *Philebus* 16C7: the gift of Prometheus is dialectic, i.e. metaphysics here.

21. cf. Plato, *Tim.* 27D-28A.
22. Plato, *Tim.* 30B; 35A-36B.
23. Plato, *Tim.* 27D6-28A1.
24. Plato, *Rep.* 509D-511E.
25. Aristotle, *Metaph.* 9.10, 1051b24-5.
26. Plato, *Phaedrus* 247C7-8.
27. i.e. the metaphysician.
28. Plato, *Soph.* 254Dff.
29. Plato, *Parm.* 129D8.
30. Plato, *Parm.* 142Bff.
31. cf. above n. 16.
32. The 'demonic' Aristotle (*daimonios*) is contrasted with 'divine' philosophers (such as Pythagoras or Plato) in later Neoplatonism. This has to do with the lower ranking of Aristotle's philosophical level (corresponding to the level of demons) as compared to the higher level of true Pythagorean-Platonists (corresponding to gods). This difference in level appears to relate to the differing quality of these philosophers' souls; cf. D. O'Meara, 1989, 37-9.
33. Aristotle, *Topics* 1.1.
34. On the relation between the metaphysician (first philosopher), the dialectician and the sophist, see Aristotle, *Metaph.* 4.2, 1004b17-26 and Syrianus' discussion of this passage below at 63,10ff.
35. i.e. the metaphysician.
36. Usener thinks there is matter missing here after *endoxôs* (compare the following, where the association of the dialectician with opinion contrasts with the association of the sophist with imagination; see Aristotle, *Metaph.* 1004b17-26).
37. i.e. dialectics and sophistry.
38. In the *Parmenides* and in the *Sophist.*
39. A word or two is missing here; my translation follows Kroll's suggested reading (*stasin <katekhein to>*).
40. This passage (6,36-7,36) is excerpted by the Byzantine philosopher Michael Psellos (*Philosophica minora* 2, 72,15-73,16), which allows us to improve the Greek text of Syrianus as transmitted by MSS later than Psellos.
41. <to be the elements> I supply from Psellos' excerpts.
42. i.e. *ta enhuparkhonta*, 'what are present in a thing', the constituent parts (components), translating with the help of Psellos' excerpts, which suggest also adding <nor> in the following line.
43. i.e. genera which are principles are neither the constitutent parts of a thing (components), nor universals abstracted *a posteriori* from perception of already constituted things.
44. Added from Psellos, *Philos. min.* 2,73,3.
45. Plato, *Philebus* 64C1. The Good is the final cause.
46. *ek tôn* added by Kroll.
47. Incorrectly.
48. i.e. universals conceived by abstraction (the *husterogenê* mentioned above).
49. i.e. individual particulars in Aristotle's *Categories*, ch. 5.
50. i.e. proximate species.
51. 'It': i.e. 'what comes after' (in the preceding line), a subordinate or partial cause (cf. below at 10,30)?
52. *mê* added by Kroll.
53. i.e. the material cause.
54. Plotinus, *Enneads* 5.7 and 5.9.
55. cf. below 38,38-9 (Iamblichus' *Commmentary on the Parmenides*).
56. cf. 103,15ff.

57. i.e. letters.

58. Syrianus suggests Aristotle's term *logois* is ambiguous, meaning either reason-principles (in this part of Syrianus' comments) or statements (in the next part).

59. The distinction between determinate in number and determinate in kind.

60. Principles composing a thing and present in it, form and matter. Syrianus distinguishes here between principles differing in kind as being immanent in a thing (form and matter), or as transcending it (efficient and final causes).

61. *epi de* is the reading in MS C; we suggest correcting this to *epi<ta>de.*

62. On these numbers, see the Introduction (section 2) to our translation of Syrianus, *On Aristotle Metaphysics 13-14.*

63. MS C has '14', a scribal mistake. Usener refers to book 12, 7-8 (see also below n. 74).

64. The determinate in kind and number. For the following compare a similar passage at 111,20-7.

65. There is a lacuna here; the translation suggests a way of supplying what is missing.

66. This is Usener's correction of the manuscript reading *praktea*; see 175,4.

67. This is the standard list of 'Pythagorean' principles; see 165,35ff.; 175,4, where references are given.

68. Syrianus may be alluding to the symbolic or mythic mode of exposition of divine principles which Proclus finds, for example, in Orphism and to the exposition of these principles by means of images which Proclus finds in Pythagoreanism; cf. Proclus, *Platonic Theology* 1,4,29,1ff.

69. See n. 60 above.

70. *ei* added by Brandis.

71. Aristotle, *Physics* 2.2, 194b13.

72. Syrianus is thinking of Aristotle's unmoved movers, as the following lines suggest.

73. i.e. Plato's philosophy.

74. MS C has '14', corrected in the edition to 12 (cf. also above 9,28). See Aristotle, *Metaph.* 12, chs 8 and 10.

75. See above n. 32.

76. *monon* is Usener's and Kroll's correction of *mallon* in the MS.

77. i.e. the mathematical objects we produce from the substantial principles immanent in soul.

78. See below 22,11-12 and 22,33-23,7; 29,13-21; 36,2-3; 40,9-10, etc.

79. We number the problems according to the order in which they are presented first by Aristotle in ch. 1. Syrianus will point out when Aristotle does not follow this order in his elaboration of the problems in the following chapters.

80. As of chapter 2, Aristotle develops the problems presented in chapter 1 by elaborating opposed argumentations for each problem. We will mark the opposition of these two phases of argumentation, to the extent that they manifest themselves clearly in Syrianus' commentary, by inserting the letters [A] and [B]. With one argumentation Syrianus will agree, rejecting that opposed to it in Aristotle's text.

81. cf. Alexander 181,2-5.

82. i.e. the second premise here, containing the major term.

83. i.e. first philosophy (metaphysics).

84. Syrianus here comments on the lines which in Aristotle follow the lemma quoted above, i.e. 996a29-b1.

85. cf. Alexander 183,9-10.

86. cf. Plato, *Theaetetus* 176B, and above, Introduction, p. 3.

87. It seems because actions (*praxeis*) are subordinate, as goods, to knowledge (*theôria*).

88. On the problem of the scientific status of mathematics, a problem raised in Plato's *Republic*, cf. Proclus, *Commentary on Euclid* 29,14ff.

89. Plato, *Rep.* 533C3-5.

90. cf. Proclus, *Commentary on Euclid* 146,24-155,23.

91. cf. 996a32.

92. Syrianus is inspired by Aristotle, *Metaph.* 13.3, 1078a36-b2 (see also Syrianus' comments on this passage at 100,15ff.). But where did he get the example of culinary art?

93. In ch. 2.

94. Alexander 184,15 refers to *Metaph.* 1 (presumably ch. 2).

95. i.e. wisdom.

96. i.e. wisdom.

97. I suggest taking *hêmas* (16,28) with *hikanon eipein*, as it makes little sense where it stands in the text, a few words later.

98. A fictitious respondent? (See the imagined discussion in what follows, where I add quotation marks for such imagined discussion.) Usener suggests reading *phusei* ('by nature') for *phêsin* ('he says'), perhaps correctly.

99. Is Syrianus thinking in general terms of Plato's *Republic* and *Laws*?

100. cf. Aristotle, *Prior Analytics* 2.2, 90a15-18; Alexander 186,1-2.

101. *Posterior Analytics* 2.10.

102. *Metaph.* 996b20-1.

103. Alexander 186,31-3 reads *ouk allês*, a reading not accepted in Jaeger's edition of the *Metaphysics*.

104. i.e. common opinions concerning wisdom (cf. *Metaph.* 1.2).

105. i.e. according to the order in which the problems are presented in ch. 1.

106. On common notions in later Neoplatonism, see above, Introduction, pp. 7-8; H. Saffrey and L. Westerink (eds) 1968, 159-61.

107. Might this perhaps be a reference to Christians?

108. *Metaph.* 1.1, 980a1.

109. For further discussion of this see below, 66,27ff.

110. i.e. the One. Cf. Plato, *Parm.* 142A, *Ep.* 7, 341C.

111. *skholê*: a course of lectures on Plato's *Parmenides*? See for example Proclus' *Commentary on the Parmenides*.

112. See above n. 82.

113. There is a lacuna in the text after these words. Syrianus will show that the principles of demonstration are *not known* in like manner by different sciences: the mathematical sciences know fundamental axioms on the basis of belief, having received them from a higher science, wisdom, which itself knows them in another, higher way, above demonstration. Cf. above n. 16.

114. i.e. first philosophy.

115. Kroll thinks that *apodeiktikên* ('demonstrative') should be added here, where there is a space in MS C.

116. cf. Plotinus 5.8.7, 38-47.

117. Reading *kata panta*, as suggested by Usener; MS C reads *katatakta*.

118. Reading *tou ... epeskemmenou* as suggested by Usener; Kroll keeps *to ... epeskemmenon*.

119. i.e. the first philosopher.

120. See above 15,7ff.

121. Reading *noêtou* with Bagolinus and Kroll; C has *nou*.

122. In ch. 1 (part 1 of Syrianus' commentary).

123. Inserted by Usener.

124. See 1003b33ff.
125. MS C has *to hoti* here (as does the following commentary); Jaeger prints *ho*.
126. Syrianus reads here the verb in the future; cf. Jaeger's apparatus to his edition of Aristotle's *Metaphysics*.
127. In chapter 1.
128. *pankhalepon* in MS C and in Aristotle; Kroll mistakenly prints *khalepon*.
129. On the distinction between belief (*pistis*) and proof, cf. above n. 16.
130. This problem and Syrianus' commentary on it are discussed by L. Cardullo, 2003, 180-208.
131. In chapter 1.
132. cf. above 3,37.
133. cf. Alexander 196,20-2; Asclepius 4,17-24.
134. MS C has *poiousin*; Aristotle in Jaeger's edition has *epoioun*, which however appears to be inserted above *poiousin* in MS C.
135. The addition of <*di'*> is suggested by Kroll.
136. cf. J. Opsomer, 2004, 31-50.
137. Alexander 197,4-28.
138. *Metaph*. 10.10, 1059a1-14.
139. Herodotus 4.94; cf. Cardullo, 2003, 161 n.
140. Plato, *Theaetetus* 173E6.
141. Plato, *Phaedrus* 247A4-5.
142. Reading *diexodikas* as proposed by Usener (correcting the MS reading *exodikas*); cf. the 'passages' in the quotation from the *Phaedrus* (above n. 141).
143. cf. the tripartition of reality in the Preface to Syrianus' commentary on book 13.
144. Plato, *Tim*. 30B5-6.
145. i.e. Aristotle. Compare 62,20, where Aristotle also contends ('fights', *diemakhometha*), this time with Empedocles.
146. Again Aristotle; cf. above 8,28-30.
147. Syrianus may be referring to the *Chaldaean Oracles*; cf. Michael Psellos, *Hupotupôsis* in *Philosophica minora* 2, 150,5-6 (derived from Proclus).
148. At 24,32-3, just above.
149. i.e. intellect, it seems.
150. i.e. Aristotle.
151. Plato, *Tim*. 34Bff.
152. Homer, *Iliad* 1.604.
153. Presumably Aristotle and Alexander of Aphrodisias.
154. Aristotle, *Metaph*. 12.7-10 (the relation between Aristotle's unmoved movers in which there appears to be a first unmoved mover).
155. Syrianus manages to find in Aristotle's astronomical system in *Metaph*. 12 the Pythagorean theory of the music of the spheres.
156. Added by Usener.
157. On 'physical numbers' in Iamblichus and Syrianus, cf. my *Pythagoras Revived*, 1989, 62 and 134.
158. Usener notes a lacuna in the text; the translation follows Kroll's suggested restitution of what is missing.
159. cf. Plato, *Rep*. 521C6ff.
160. On the basis of an inborn common notion, better known than what is demonstrated? Cf. 76,34-5.
161. Aristotle compares nature with a self-healing doctor in *Physics* 2.8, 199b26-33 and 2.1, 192b23-7 (we owe these references to Euree Song).

162. For the comparison of justice with medicine, cf. Plato, *Gorg.* 464B-C; both correct dysfunctions, the former those of the soul, the latter those of the body.

163. Their models?

164. cf. below 38,37-39,6 (where references are given).

165. MS C has *oud' hê peri ton ouranon astrologia...*, whereas Jaeger prints *oude peri ton ouranon hê astrologia...* (see his apparatus).

166. cf. the making of the world-soul and of human souls in Plato's *Timaeus*. There is a similar passage further on in Syrianus' commentary, at 82,29-36.

167. i.e. soul, as *demiurgic, making* the world according to the reason-principle, or simply (in the following words of Syrianus) soul, as *human* soul, just *knowing* the reason-principle.

168. For the argument for the priority of mathematicals in relation to perceptibles, cf. 95,29ff., Proclus, *in Eucl.* 12,9ff.

169. *kanôn*, cf. the lemma; perhaps the sense is wider here: criterion?

170. Added by Usener.

171. Following Kroll's suggested correction of *ta* to *tôn*.

172. I keep the *pros* of MS C, rather than the correction to *pro* suggested in Kroll's edition. The point here is that the common is in the particular, not that it is (also in some sense) prior (a point made later at 29,3ff.).

173. On the Stoic *idiôs poion* (the 'peculiarly qualified'), cf. A. Long and D. Sedley, *The Hellenistic Philosophers*, 1987, ch. 28. Long and Sedley give Syrianus 29,18-19 as passage 28G (they follow Kroll's correction *pro*; cf. preceding note).

174. i.e. universals in particulars, what is common in particulars.

175. MS C has *phônai pasai*; Jaeger does not print *pasai*.

176. MS C has *esti* here, where Jaeger prints *sunestêke*

177. cf. above 5,8ff.

178. cf. above, 7,7-21.

179. *epigennêmatika*, i.e. as concepts derived by abstraction.

180. MS C: *kai*; Jaeger has *k'an*.

181. See above 6,36-7,21.

182. MS C has *esti*; Jaeger prints *estai*.

183. Aristotle, *Posterior Analytics* 2.10, 93b35-7; 2.13. Alexander (204,13-14) may however be the source of Syrianus' reference (in his edition of Alexander, Hayduck corrects the reference to *Topics* 6.4).

184. Reading *genikai* as suggested by Usener.

185. cf. 6,36-7,21.

186. As presented in ch. 1.

187. cf. 7,21.

188. MS C has *hoti* (perhaps a scribal error) for Aristotle's *aei*.

189. This represents the correct position, in Syrianus' view; see below 35,21ff.

190. *epigennêmatika*: i.e. genera as abstracted universals, what Syrianus also calls 'later-born' (*husterogenê*); cf. above at 7,10ff.

191. MS C has *gar*: Jaeger's text of Aristotle has *de*.

192. Not marked as a lemma in Kroll.

193. Not presumably the Pythagoreans and Platonists, but those alleged by Aristotle to hold the positions he refutes.

194. Alexander 206,22ff.

195. Added by Kroll.

196. MS C has *hôs* here, not printed by Jaeger.

197. There is a gap here in MS C (... *kai tou*) , for which Usener proposes *touto kata tou*.

198. With Plato, *Tim.* 28A4-6.

199. Following Usener's suggestion to add *genos* here.

200. Following Kroll's suggestion to add *ton de sullogismon* here.
201. See n. 82.
202. This is a kind of ordered series described as a 'P-series' by A. Lloyd, *The Anatomy of Neoplatonism*, Oxford 1990, 76-8.
203. On reversal in arguments (*peritropê*), a Sceptic technique, in later Neoplatonism, see A. Linguiti, 1990, 68-73.
204. cf. Plotinus, *Enneads* 1.8.14, 44-8.
205. I suggest reading *ei* here for *kai* (as referring to the hypothesis which is the basis of the argument).
206. Reading *tauta* with Usener; Kroll prints *taûta*.
207. cf. *Categories*, ch. 5.
208. This problem and Syrianus' commentary are discussed by Cardullo, 2003, 208-22.
209. In book 3, ch. 1.
210. *kai* inserted by Kroll.
211. Following Kroll's suggested correction of the text, which is corrupt here.
212. i.e. the heavenly bodies.
213. The celestial bodies are secondary causes, contributing to the coming-to-be of things.
214. i.e. not conceptual abstractions.
215. MS C has *ei esti eidos ti*; Jaeger's text of Aristotle just has *ei esti*.
216. Alexander 211,34.
217. The demiurge of Plato's *Timaeus*, it would seem.
218. cf. Plato, *Theaetetus* 151E2-152C6.
219. These include in particular mathematical objects.
220. This is a favourite Platonic passage for Syrianus; cf. also 41,11-12; 118,6-7.
221. cf. Aristotle, *On the Heavens*, 1.5; Proclus, *Elements of Physics* 2.15.
222. cf. above n. 219.
223. i.e. in the presentation of the problems in ch. 1.
224. i.e. in the discussion of the problems, just above, 37,25.
225. *hoi amphi ton Parmenidên*; cf. the first part of Plato's *Parmenides*.
226. Iamblichus, *Commentary on the Parmenides*, fr. 1 Dillon (the present passage).
227. Plotinus, *Enneads* 5.9, 10-14.
228. cf. also 107,5-108,5 for a fuller discussion.
229. i.e. enmattered form.
230. This is supplied by Usener.
231. *ou gar pantôn hê ousia mia*; MSS of Aristotle (see Jaeger's apparatus) and Alexander's lemma (see the Dooley-Madigan translation, 161 n. 322) have *ou gar hen hapanta hôn hê ousia mia*.
232. i.e. in the demiurgic intellect.
233. i.e. with matter.
234. *Tim.* 50C4-6.
235. *Parm.* 131A-C, summarised in the following lines in Syrianus.
236. Added by Kroll.
237. *proteinôn*: Syrianus uses the same verb in speaking of Aristotle's presentation of problems in *Metaph.* 3 (for example at 1,20), which suggests an analogy between the metaphysical 'exercises' of *Metaph.* 3 and those of the *Parmenides* (see Introduction, p. 8).
238. Kroll refers to Plato, *Sophist* 248A, but the reference may be more general. Cf. Syrianus at 83,34-5; 120,17-18.
239. See 108,31ff.
240. Kroll refers to *Metaph.* 8.6.

241. cf. above at 29,13ff.
242. I punctuate (with Schneider) differently from Kroll here.
243. 'up to' indicates in MS C that the following part of the lemma (supplied here in square brackets) is omitted.
244. 9,1ff.
245. As presented in ch. 1.
246. *aphtharta* MS C, corrected by Kroll to *pôs phtharta*.
247. *aidia*, Kroll's correction of *aidiai*.
248. An echo of *Phaedrus* 245E1; cf. above at 37,14-15; 44,22.
249. See for example Proclus, *Elements of Theology*, prop. 76.
250. Kroll substitutes the text of the *Timaeus* for *hin' oun thnêton te kai athanaton deontôs hapan* in Syrianus' quotation (which is translated here).
251. The language here recalls Empedocles (cf. fr. 115), who will soon be cited.
252. On this cf. I. Hadot, 1989, 113-22.
253. cf. the four ways (including the 'inspired' way, *entheastikôs*) in which theology is communicated in Proclus, *Platonic Theology*, 1,4.
254. cf. Plato, *Tim.* 40E1-2.
255. Kroll refers to Plato, *Laws* 629B, 682A, *Ion* 530B, but see especially *Tim.* 40D6-E4.
256. In Plato's *Timaeus* as interpreted in Neoplatonism; cf. above 41,13ff.
257. Above, 41,16 (citing Plato, *Tim.* 41B8).
258. See Proclus, *Elements of Theology*, prop. 76.
259. Empedocles fr. 30,2, quoted by Aristotle at 1000b15.
260. Orpheus fr. 68.
261. Empedocles fr. 30,3, quoted by Aristotle at 1000b15a (cited here a little later).
262. Orpheus fr. 66.
263. Following Kroll's correction of *anelabeto* to *apelaueto*.
264. Empedocles fr. 109,1, quoted by Aristotle 1000b6a.
265. The text is corrupt here: the translation suggests what seems to be the sense (cf. *Metaph.* 12.9, 1074b32; Alexander 220,26-7).
266. Added by Usener.
267. Above at 11,25ff.
268. Added by Usener.
269. Added by Brandis.
270. This is the language of Plato's *Parmenides* read as indicating a metaphysical hierarchy; cf., for example, Plotinus 5.1.8, 25-6.
271. Something like this must be supplied.
272. Translating *eti* with Aristotle; MS C has *hoti*.
273. On substantial number see our Introduction (section 2) to *Syrianus: On Aristotle Metaphysics 13-14*.
274. Supplying *katholou* from Aristotle's text and from Syrianus 46,10; *kath' hou* is printed by Jaeger.
275. cf., for example, Plotinus 5.1.6, 3ff.
276. cf. Parmenides fr. 8,13.
277. cf. Parmenides fr. 8,23-4.
278. Added by Usener.
279. MS C has *gar kai*; Jaeger prints *gar*.
280. *anankê de mêden einai* MS C; cf. 47,15-17. Jaeger prints *anankê gar mê hen einai*.
281. Or Plato?
282. cf. Alexander 227,21-2.
283. Following Kroll's correction of *hênômenon* to *hidrumenon*.

284. I translate following Bagolinus's suggestion (in Kroll) to suppress *pros* at 48,5.

285. Plato, *Tim.* 27C3.

286. Plato, *Rep.* 508B13.

287. This problem in Aristotle and the commentaries on it of Alexander, Syrianus and Asclepius are analysed and compared by F. de Haas, 1997, 347-68.

288. Above 44,30.

289. Above 12,25ff.

290. cf. *Metaph.* 5.11, 1019a1-11 on Platonic priority in substance (Syrianus gives examples of this principle in the following lines).

291. For the distinction, cf. already Plotinus, *Enneads* 5.6.3, 10-15.

292. *aülotera*, Kroll's correction of *atelestera.*

293. cf. Alexander 230,30-2.

294. Usener's addition.

295. Usener's conjecture, reading *toiouto* for *touto.*

296. 'not' is suppressed by Kroll.

297. MS C has *energeiai;* Kroll suggests reading *tina;* Usener thinks that *diaireseis* has been omitted. What seems indeed to be meant is the division of a body, e.g. surface.

298. i.e. point, line, surface.

299. MS C has *alla toiauta;* Jaeger prints *all'atta.*

300. See above 40,9ff.

301. Translating following Usener's suggestion to supply these words here.

302. Added by Usener.

303. Plato, *Rep.* 475E4.

304. Syrianus means probably *Metaph.* 12.6-7.

305. Following Usener's suggested way of supplementing what is missing in the text.

306. Knowledge derives from innate *logoi* (mentioned just before, at 53,5), not from universals produced *a posteriori* by abstraction.

307. i.e. Aristotle's unmoved movers?

308. Alexander 236,26-8.

309. Above 29,13ff.

310. In the fifth problem.

311. cf. 1,8-12; 74,30; Alexander 238,1-3.

312. Usener supplies <*estin epistêmês eidenai*>, cf. 57,28; Kroll adds <*tês*> (i.e.: '... it belongs to one science, the science of being as being, to know the causes').

313. Supplied by Usener.

314. Alexander of Aphrodisias; on Syrianus' use of Alexander, see Introduction, section 4.

315. Translating Usener's suggestion to add *axion* here.

316. As in the translation of Syrianus' commentary on *Metaph.* 3, we have added here to our translation the division into chapters of Aristotle's text and some titles for sections.

317. *tôn allôn;* in Aristotle: *tôn en merei legomenôn.*

318. *toutou;* Jaeger prints *touto.*

319. This quotation is marked as a lemma in MS C, but it seems better to treat it as a quotation here.

320. cf. 1003a23-6.

321. cf. Plato, *Rep.* 509B9.

322. Platonists.

323. cf. Proclus, *Elements of Theology*, prop. 67.

324. cf. 1003a26-8.

325. cf. 1003a28-9.
326. cf. 1027a7-8.
327. *Metaph.* 5.7; 7.1, 1028a10.
328. *paidion* (more likely 'slave' here than 'small child').
329. cf. *Nicomachean Ethics* 1.4, 1096a13-27.
330. Supplied by Usener.
331. Reading *gennêtika* with Aristotle (1003b8) and Alexander (242,27); cf. Syrianus 57,1. Kroll prints *poiêta*, MS C has *poiêtika*.
332. Kroll conjectures *aisthomenoi*, MS C has *ekthem* with *ous* or *ês*. On the theory cf. Aristotle, *On Generation of Animals* 3.11; 5, 4.
333. *aphthonôs*: cf. Plato, *Timaeus* 29E1-2 (the generosity of the divine).
334. Above, 54,5.
335. cf. 54,4-8.
336. In the next section, 58,24ff.
337. cf. 1003b19-26.
338. Usener and Kroll correct the MS reading *grammatikos* to *arithmetikos*; the correction is supported by the context and by Nicomachus, *Introduction to Arithmetic* 1.3; Iamblichus, *On Common mathematical Science*, 30,3-25. The example of the *grammatikos* in Aristotle (1003b20) may have influenced the manuscript tradition of Syrianus.
339. cf. 1003b21-2; Alexander 245,37ff.
340. MS C has 'second' (*deuterô*), derived incorrectly from '*B* ' (noted by Luna, 2001, 90n.); cf. Alexander 246,13-15.
341. cf. 986a23-6, 1004b27.
342. cf. Alexander 247,22.
343. Syrianus contrasts those (Platonists) who are concerned with the truth of the matter (if being is the same as unity) with those (Aristotelians?) whose concern is finding consistency in what Aristotle says about the relation between being and unity. He will argue against the latter.
344. cf. Proclus, *Elements of Theology*, props. 1-4.
345. Reading *huperousiou* (MS C : *huperousia*; Kroll: *huper ousian*).
346. Deleting *mê* with Kroll.
347. Reading *holôs* with Kroll; MS C: *homôs*.
348. Added by Kroll, as is <unity>; perhaps one should read 'if unity is substance', as suggested by Usener. The following argument is dialectical: if Aristotle says that unity is the same as being, then he will have to conclude that numbers in material things (things numbered) are substances, a position he could not accept. So he must concede that unity and being are not the same in material things.
349. cf. Plotinus, *Enneads* 6.9.1, 1.
350. See above, commentary on *Metaph.* 3, n. 32.
351. cf. above n. 341.
352. Kroll marks a lacuna here, but nothing seems to be missing.
353. On these columns, cf. above n. 341.
354. Aristotle refers to this (lost) work at 1004a2.
355. cf. 1004a2-3.
356. cf. 1004a10-12.
357. cf. Alexander 253,8ff.
358. <*ennoian*> supplied by Usener.
359. cf. Alexander 254,10-12.
360. 'We' and 'you' here in this section refer ironically to Aristotle.
361. cf. 59,7, on Aristotle's inconsistency.
362. The fifth puzzle of book 3 (995b19-22).

363. The first philosopher, or metaphysician.
364. Added by Usener from Aristotle.
365. Nicomachus *Introduction to Arithmetic*, 1.17; Iamblichus, *Commentary on Nicomachus' Introduction*, 37,1.
366. cf. 1004b17-18.
367. i.e. the dialecticians.
368. Added by Brandis.
369. On the perfect philosopher, cf. above commentary on *Metaph.* 3, n. 11.
370. Usener corrects *sunônumôs* in MS C to *homônumôs*. However the MS reading should be kept, as suggested by Schneider. Unity is not univocal, according to Syrianus, but equivocal in a certain sense (by reference to one thing).
371. Should 'perfect' (*teleiou*) be corrected to 'equal' (*isou*)?
372. cf. Alexander 264,11-12.
373. i.e. multiplicity.
374. I translate Syrianus as contrasting here those Pythagoreans who start with a pair of opposed first principles (see Aristotle, *Metaph.* 1.5, 986a18-20) with 'better' Pythagoreans who have *one* first principle transcending the pair; see Syrianus 112,14ff.; Sextus Empiricus, *Adv. math.* 10.282.
375. i.e. the first pair in the two columns: limit and the unlimited.
376. Euclid, *Elements* 1, common notion 1, considered by Syrianus as an axiom proper to geometry and, as such, a principle of demonstration specific to geometry.
377. Euclid, common notion 3.
378. i.e. the first philosopher, or metaphysician.
379. cf. 1005b2-4 ('analytics': cf. *An. Post.* 1.2-3).
380. See below at 65,32-4.
381. Syrianus distinguishes between three formulations of the Principle of Non-Contradiction, which do not represent three autonomous principles, as A. Longo argues in her discussion of this passage (2005, 119-27; see above, Introduction, p. 7), but correspond, I would suggest, to three levels on which the Principle is expressed (see the next paragraph in Syrianus): in things (*pragmata*), in discursive dispositions of the soul (*ennoêmata*), and in propositions (*logoi*).
382. cf. 1005b29.
383. Ch. 14; cf. Alexander 270,24-5.
384. Our title gives the way in which this principle is normally designated today. However, Syrianus himself refers to the principle as: 'contradiction' (*antiphasis*), or 'the axiom of contradiction' (*to tês antiphaseôs axiôma*).
385. We have added some sub-titles so as to ventilate a little the following very dense and complex discussion (as dense and complex indeed as the corresponding discussion in Aristotle).
386. cf. 1006a5-11; Alexander 272,24-6.
387. cf. Plato, *Phaedrus* 245D1-E2 , quoted above at 37,14-15 (with our note 220).
388. Added by Usener. In the following example, Syrianus argues that the denial of an axiom specific to mathematics can be reduced to the denial of the principle of non-contradiction.
389. Euclid, *Elements* I, common notion 3.
390. i.e. to axioms such as the more specialised axioms in geometry formulated in Euclid's common notions (above nn. 376, 377).
391. cf. 1006a11-15.
392. cf. 1006a15ff.
393. cf. 1006b7.
394. cf. 1006a34.

395. MS C has *mon* and Kroll prints *mousikon* with Aristotle (1006b16).

396. Replacing *nemontos* with *onoma* as suggested by Kroll.

397. i.e. to the claim that the negation is false of that of which the affirmation is true (cf. 67,15-16).

398. i.e. that that of which the affirmation is true, the negation is false.

399. Syrianus takes the two arguments of the objector in reverse order, first the supposition [i] that all flows, then [ii] that things have a stable and determinate nature.

400. cf. 1010a11-13.

401. Syrianus returns to the position argued at 67,3ff. that a name signifies one thing.

402. Already argued at 66,35-67,3.

403. i.e. both affirmation and negation as parts of a contradiction.

404. cf. 65,35-66,5.

405. cf. 1006a12.

406. Alexander 273,18-19 (= Fragment 85 in W. Fortenbaugh *et al.*, *Theophrastus Fragments*, Leiden 1990).

407. cf. 1006b34-1007a1.

408. Supplied by Usener and accepted by Kroll.

409. cf. 1007a8-20.

410. On this cf. Alexander 284,12ff.

411. *Posterior Analytics* 1.4, 73b26ff.; *Metaph.* 1007a34; cf. Alexander 290,1-10.

412. *paidarion*: perhaps 'slave'.

413. Added by Usener.

414. Syrianus finds this claim in Aristotle 1007b1-2 (*oude gar pleiô sumpleketai duoin*); cf. Alexander (288,29-289,20) and Asclepius 265,19-22, who reports Ammonius' critique of this.

415. cf. above n. 414.

416. On this passage see the discussion in Longo, 2005, 369-70.

417. cf. *Posterior Analytics* 2.3 and 10.

418. Keeping *horôsi* with MS C; Kroll prints *eôsi*.

419. cf. 1007b22-3; 1009a6-8.

420. 1007b25-6.

421. *to heteron ... tôn tês antiphaseôs axiômatôn*, i.e. the Principle of excluded middle.

422. The language suggests the technique of reversal (*peritropê*) used in Scepticism and in Neoplatonism; cf. A. Linguiti, 1990, 68-73.

423. cf. 1008a15-16.

424. Reading *didoasi ti kai hôristhai* with Usener; Kroll prints *didoasi te kai hôristai*.

425. cf. 1008a7-11.

426. MS C has: *helôntai kai koinêi tôn haireseôn hekatera kai idia*, corrected by Usener, who is followed by Kroll, who prints however the MS reading *haireseôn* (whereas Usener suggests *theseôn*) and suggests instead *diaireseôn* (from Alexander). The two positions (which I distinguish with [i] and [ii]) are: [i] that each member of a pair of contradictory statements, by itself, is true, and [ii] that both members, taken together, are true (cf. 1008a7-8).

427. cf. 1008a22-3.

428. cf. 1008b2-3.

429. Added by Kroll from Aristotle.

430. cf. 1008b11-12.

431. In one's soul; cf. above 65,22.

432. cf. 1008b14ff.

433. On the 'Ephectics' (i.e. Sceptics) in later Neoplatonic texts, cf. I. Hadot et al., 1989, 57-60; H. Flückiger, 2005, 127.
434. cf. 1008b28ff.
435. i.e. the transcendent objects (mathematical objects and Forms) of Platonism.
436. See above commentary on *Metaph.* 3, n. 106, and below 76,33-5.
437. Reading *diathrêsas* with Usener; MS C has *dia thêras* which Kroll corrects to *diakathêras*.
438. cf. Alexander 303,25-8.
439. Kroll suggests that some words are missing here, which would have referred to followers of Anaxagoras, given the reference which follows to homoeomeries; see Aristotle 1009a27.
440. cf. 1009a34-6; 1017a35-1017b1; Alexander 304,13-16.
441. cf. 1009a37.
442. cf. 1009b2ff. Here and in what follows Syrianus evokes *Aristotle's* reports in the text on a series of Presocratic authors, reports with which Syrianus does not necessarily agree. We have therefore given references to Aristotle's reports, rather than to the original works of these authors.
443. Added by Brandis.
444. cf. 1009b11-12.
445. cf. 1009b17-25.
446. cf. 1009b26-8.
447. Reading *homôs* as suggested by Kroll; MS C has *homoiôs*.
448. 1009b28-30.
449. cf. 1009b33-1010a1.
450. 1010a13-15.
451. cf. Alexander 310,18-19.
452. 1010a35-6.
453. 1010a37-1010b1.
454. cf. 1010b1ff.
455. cf. above commentary on *Metaph.* 3, n. 82.
456. cf. 1010b11; and above 73,10.
457. cf. 18,10ff.
458. cf. 1010b11-14.
459. cf. 1010b14-19.
460. cf. 1010b30-1011a2; Alexander 316,13ff.
461. Usener suggests reading *<ta> panta kat' autous <ei> aisthêta* ('all things will not exist, <if>, for these people, they are perceptibles').
462. In this case, perceptibles as objects of perception.
463. i.e. the perceptive faculty.
464. cf. above, 74,28-31.
465. cf. 74,26-8.
466. cf. 1011a15-16.
467. cf. above n. 422.
468. *prosêkanto*; Kroll prints *prosêlanto*.
469. Plato, *Theaetetus* 165E2.
470. Kroll notes a gap here in MS C, but nothing seems to be missing.
471. cf. 1011b5-7.
472. Euripides, *Medea* 1417-18.
473. cf. 1011b9-12.
474. cf. above 65,21-2.
475. The excluded middle.
476. cf. 1011b27.

477. cf. 1011b26.

478. cf. 1011b30-1.

479. cf. 1012b13ff.

480. cf. 1012b23ff.

481. cf. above 11,2-5.

482. Kroll adds '<not>' here, unnecessarily: the acceptability of this statement, as equivocal, parallels that of the preceding example given, the statement (also equivocal) that 'all things desire the good'.

Select Bibliography

The following is a list of the principal works cited or utilised in the introduction and notes. See also the Select Bibliography given in *Syrianus: On Aristotle Metaphysics 13-14,* tr. J. Dillon & D. O'Meara, mentioned below.

Primary sources

Alexander Aphrodisiensis, *In Aristotelis metaphysica commentaria,* ed. M. Hayduck, Berlin: Reimer, 1881 (*Commentaria in Aristotelem Graeca* 1: 1).

Alexander of Aphrodisias, *On Aristotle Metaphysics 2 & 3,* tr. W. Dooley & A. Madigan, London: Duckworth, 1992.

Alexander of Aphrodisias, *On Aristotle Metaphysics 4,* tr. A. Madigan, London: Duckworth, 1993.

[Anonymous] *Prolegomena to the Philosophy of Plato,* ed. L.G. Westerink, J. Trouillard & A. Segonds, *Prolégomènes à la philosophie de Platon,* Paris: Les Belles Lettres, 1990.

Asclepius, *In Aristotelis metaphysica commentaria,* ed. M. Hayduck, Berlin: Reimer, 1888 (*Commentaria in Aristotelem Graeca* 6: 2).

Iamblichus, *In Platonis dialogos commentariorum fragmenta,* ed. J. Dillon, Leiden: Brill, 1973.

Iamblichus, *Protrepticus,* ed. L. Pistelli, Stuttgart: Teubner, 1888.

Iamblichus, *De communi mathematica scientia,* ed. N. Festa, Leipzig: Teubner, 1891.

Iamblichus, *In Nicomachi arithmeticam introductionem liber,* ed. H. Pistelli, Stuttgart: Teubner, 1975.

Marinus, *Vita Procli,* ed. H.-D. Saffrey & A. Segonds, *Marinus Proclus ou sur le bonheur,* Paris: Les Belles Lettres, 2001.

Orphicorum fragmenta, ed. O. Kern, Berlin: Weidmann, 1922.

Proclus, *Elements of Physics,* ed. A. Ritzenfeld, Leipzig: Teubner, 1912.

Proclus, *Elements of Theology,* ed. E.R. Dodds, Oxford: Clarendon Press, 1963.

Proclus, *In primum Euclidis Elementorum librum commentarii,* ed. G. Friedlein, Leipzig: Teubner, 1873.

Proclus, *Platonic Theology,* ed. H.-D. Saffrey & L.G. Westerink, *Proclus Théologie Platonicienne,* Paris: Les Belles Lettres, 1968-1997.

Psellos, *Philosophica minora,* vol. 2, ed. D. O'Meara, Leipzig : Teubner, 1989.

Syriani antiquissimi philosophi interpretis in II, XIII, et XIV libros Aristotelis Metaphysices commentarius, a H. Bagolino latinitate donatus, Venice, 1558.

Syrianus, *In Aristotelis Metaphysica commentaria,* ed. H. Usener, Berlin: Reimer, 1870.

Syrianus, *In Metaphysica commentaria,* ed. W. Kroll, Berlin: Reimer, 1902 (*Commentaria in Aristotelem Graeca,* 6: 1).

Syrianus: On Aristotle Metaphysics 13-14, tr. J. Dillon & D. O'Meara, London: Duckworth, 2006.
Themistius, *Orationes*, ed. H. Schenkl & G. Downey, Leipzig: Teubner, 1965-1970.

Secondary works

Bonelli, M., *Alessandro di Afrodisia e la metafisica come scienza dimostrativa*, Naples: Bibliopolis, 2001.
Cardullo, L., 'Syrianus défenseur de Platon contre Aristote, selon le témoignage d'Asclépius (*Métaphysique* 433.9-436.6)', in *Contre Platon 1. le platonisme dévoilé*, ed. M. Dixsaut, Paris: Vrin, 1993, 197-214. (1993a)
Cardullo, L., 'Giamblico nel' *Commentario alla Metafisica* di Siriano', in *The Divine Iamblichus*, ed. H.J. Blumenthal & E.G. Clark, London: Bristol Classical Press, 1993, 173-200. (1993b)
Cardullo, L., *Siriano Esegeta di Aristotele*, Florence: La nuova Italia, 1995.
Cardullo, L., ' "Come le frecce dei Traci ..." Siriano contro Aristotele a proposito di due aporie di *metafisica* B sul soprasensibile', *Il libro B della* metafisica *di Aristotele*, ed. V. Celluprica, Naples: Bibliopolis, 2003, 161-225.
D'Ancona, C., 'Syrianus dans la tradition exégétique de la *Métaphysique* d'Aristote: Antécédents et postérité', *Le Commentaire entre tradition et innovation*, ed. M.-O. Goulet-Cazé, Paris: CNRS, 2000, 311-27.
D'Ancona, C., 'La Doctrine des principes: Syrianus comme source textuelle et doctrinale de Proclus. 1e partie: Histoire du problème', *Proclus et la théologie platonicienne*, ed. A. Segonds and C. Steel, Leuven-Paris: Leuven University Press, 2000, 189-225.
Flückiger, H., 'The *EPHEKTIKOI* in the Commentators', in A. Brancacci, ed., *Philosophy and Doxography in the Imperial Age*, Florence: Olschki, 2005, 113-29.
Haas, F. de, 'Aristotle's Fourteenth Aporia and the three dimensions in later Neoplatonism', *The Perennial Tradition of Neoplatonism*, ed. J. Cleary, Leuven: Leuven University Press, 1997, 347-68.
Hadot, I., *Simplicius Commentaire sur les Catégories*, fasc. 1, Leiden: Brill, 1989.
Kremer, K., *Der Metaphysikbegriff in den Aristoteles-Kommentaren der Ammonius-Schule*, Münster 1961.
Larsen, B., *Jamblique de Chalcis : exégète et philosophe*, Aarhus: Aarhus University Press, 1972.
Linguiti, A., *L'Ultimo Platonismo greco*, Florence: Olschki 1990.
Lloyd, A., *The Anatomy of Neoplatonism*, Oxford: Clarendon Press, 1990.
Longo, A., *Siriano e i principi della scienza*, Naples: Bibliopolis, 2005.
Longo, A., ed., *Syrianus et la métaphysique de l'antiquité tardive*, Naples: Bibliopolis (to appear).
Luna, C., 'Syrianus dans la tradition exégétique de la *Métaphysique* d'Aristote. I: Syrianus entre Alexandre d'Aphrodise et Asclépius', *Le Commentaire entre tradition et innovation*, ed. M.-O. Goulet-Cazé, Paris: CNRS, 2000, 301-9.
Luna, C., 'La doctrine des principes: Syrianus comme source textuelle et doctrinale de Proclus. 2e partie: analyse des textes', *Proclus et la théologie platonicienne*, ed. A. Segonds and C. Steel, Leuven-Paris: Leuven University Press, 2000, 227-8.
Luna, C., *Trois études sur la tradition des commentaires anciens à la* Métaphysique *d'Aristote*, Leiden: Brill, 2001.
Luna, C., 'Alessandro di Afrodisia e Siriano sul libro B della *Metafisica*: tecnica e struttura del commento', *Documenti e studi sulla tradizione filosofica medievale* 15, 2004, 39-79.

Luna, C., 'Observations sur le texte des livres M-N de la *Métaphysique* d'Aristote', *Documenti e studi sulla tradizione filosofica medievale* 16 (2005), 553-93.

Luna, C., 'Mise en page et transmission textuelle du commentaire de Syrianus sur la *Métaphysique*', *The Libraries of the Neoplatonists*, ed. C. d'Ancona, Leiden: Brill, 2007, 121-33.

O'Meara, D., 'Le Problème de la métaphysique dans l'Antiquité tardive', *Freiburger Zeitschrift für Philosophie und Theologie* XXXIII (1986), 3-22.

O'Meara, D., *Pythagoras Revived: Mathematics and Philosophy in Late Antiquity*, Oxford: Clarendon Press, 1989.

O'Meara, D., ' La science métaphysique (ou théologie) de Proclus comme exercice spirituel', *Proclus et la théologie platonicienne*, ed. A. Segonds & C. Steel, Leuven-Paris: Leuven University Press, 2000, 279-90.

O'Meara, D., *Platonopolis: Platonic Political Philosophy in late Antiquity*, Oxford: Clarendon Press, 2003.

Opsomer, J., 'Syrianus on Homonymy and Forms', in *Platonic Ideas and Concept Formation in Ancient and Medieval Thought*, ed. G. van Riel & C. Macé, Leuven: Peeters, 2004, 31-50.

Panagiota Manolea, C., *The Homeric Tradition in Syrianus*, Thessaloniki: Ant. Stamoulis Editions, 2004.

Steel, C., *The Changing Self: A Study on the Soul in Later Neoplatonism*, Bruxelles: Paleis der Academien, 1978.

Verbeke, G., 'Aristotle's Metaphysics Viewed by the Ancient Greek Commentators', *Studies in Aristotle*, ed. D. O'Meara, Washington D.C: Catholic University Press, 1981, 107-27.

English-Greek Glossary

This glossary lists a selection of the more important words, either from a philosophical or philological perspective, occurring in the Greek text. The translations given here may not always correspond to the rendering of them in a particular passage in the English text, since the demands of idiomatic translation may call for variation; but it should always be possible to work out what word is being translated.

able to say the truth: *alêtheutikos*
above substance, being: *huperousios*
absolutely good: *panagathos*
abstraction: *aphairesis*
absurd: *atopon*
abyss: *akhania*
account: *apodosis*
action: *praxis*
active: *drastêrios*
activity, actuality: *energeia*
affection, attribute: *pathos*
affection: *pathêma*
affirmation: *kataphasis*
alteration: *alloiôsis*
ambrosia: *ambrosia*
analyse: *analuein*
analytic: *analutikos*
angels: see order of angels
anhypothetical: *anupothetos*
appearance, image, imagination: *phantasia*
apprehension, focus: *epibolê*
arbitrarily, for no good reason: *eikêi*
archetypal: *arkhetupos*
argue, attempt: *epikheirein*
argument: *skemma*
argumentative exercise: *gumnastikos (logos)*
art, skill: *tekhnê*
assimilation: *exhomoiôsis*
astronomy: *astrologia*
axiom: *axiôma*

being as being: *to on hêi on*
being itself: *autoon*
being: *to on, einai*
benefit: *agathunein*
body-like: *sômatoeidês*
by derivation from, by reference to one thing: *aph' henos, pros hen*

care for: *epimelêthênai*
care for: *pronoein*
care: *epimeleia*
caring: *promêtheia*
category, predication: *katêgoria*
cause: *aitia, aition*
clarity, evidence: *enargeia*
coherence: *sunokhê*
column: *sustoikhia*
coming after, later born, *a posteriori*: *husterogenês*
coming-to-be, becoming, generation: *genesis*
common notions, conceptions: *koinai ennoiai*
common quality: *koinôs poion, to*
conception: *ennoia*
concern: *pronoêtikos*
consider, to, know: *theôrein*
consistency: *homologia*
constituents, components: *enuparkhonta, ta*
constitution, constituted nature: *sustasis*
constrict: *stenokhôrein*
contemplative, more: *theôrêtikôteros*
contradiction: *antiphasis*

contraries: *enantia*
contrariety: *enantiôsis*
co-ordinate, to: *suntassein*
co-ordinate: *katatetagmenos*
co-ordinated: *katataktos*
co-ordination, order: *suntaxis*
cut off: *apotemakhizein*

deceive: *sophizein*
declarative: *apophantikos*
declension: *apoptôsis*
definition: *horismos, dioristikos*
definitory: *horistikos*
demiurge: *dêmiourgos*
demiurgic: *dêmiourgikos*
demonic, miraculous: *daimonios*
demonstration: *apodeixis*
demonstrative: *apodeiktikos*
desired, much: *poluaratos*
destroys: *anairetikos*
destruction: *phthora*
dialectic: *dialektikos*
difference, differentia: *diaphora*
difficulty, puzzle: *aporia*
directive: *hêgemonikos*
disagree, to be discordant: *diaphônein*
discursive thinking: *dianoêsis*
discursive: *dianoêtos*
disposed, be: *diakeisthai*
disposition: *diathesis*
disputatious: *agônistikos*
dissimilar: *anomoios*
distinctive property: *idiotês*
divided in parts, as: *merikôs*
divided: *meristos*
divine science, higher in: *theologikôteros*
divine: *theios*
division: *diairetikê*
doctrine: *dogma*
double: *anadiploun*

eidetic number, form-number: *eidêtikos arithmos*
element: *stoikheion*
empty: *diakenos*
engineer: *mêkhanikos*
enmattered: *enulos*
enumerate: *aparithmêsis*
ephectic: *ephektikos*
equal: *isos*
equality: *isotês*
equator: *isêmerinos*
equivocal (adv.): *homônumôs*
especially: *prohêgoumenôs*
essential accident: *kath' hauto sumbebêkon*
essential attribute: *kath' hauto pathos*
essential number: *autarithmos*
essential property: *kath' hauto huparakhon*
essentially, essential: *kath' hauto*
estranged: *allotriotês*
eternal motion (movement), in: *aeikinêtos*
eternal movement: *aeikinêsia*
eternity: *aidiotês*
examine: *basanizein*
exercise before: *progumnazein*
exercise, practise: *gumnazein*
exercise: *gumnasia*
existence, nature: *hupostasis*
existence: *huparxis*
explicate: *anelissein*
exploration: *diaporia*
exploring difficulties: *diaporêtikos*
expository: *huphêgêmatikos*
express: *hermêneuein*

faculty of judgement: *kriterion*
fair-mindedness: *epieikeia*
false beliefs: *pseudodoxia*
figure: *skhêma*
final: *telikos*
fine, beautiful: *kalos*
firm: *bebaios*
first philosophy, he who practises: *philosophos, ho prôtos*
flavour: *khumos*
Form (Platonic): *idea*
Form, species, kind, part: *eidos*
function: *ergon*

general: *holikos, genikos*
generated additionally: *epigennêmatikos,*
goal: *telos*
god: *theos*
good, the: *to agathon*
goodness: *agathotês*
grasp: *katanoêsis*

harmonic: *harmonikos*
health itself: *autohugieinon, to*
hearth: *hestia*
heavens, the: *ouranos*

henad: *henas*
hint, enigmatic: *ainittesthai*
homoeomerous reality: *homoiomereia*
hypothesis: *hupothesis*

image: *eidôlon, indalma, eikôn*
imaginative, the: *phantastikon, to*
imagine: *periphantazesthai*
imagining: *eikotologia*
immaterial: *aülos*
immateriality: *aülia*
immediate premise: *amesos protasis*
immediately, self-evident: *autothen*
impassible: *apathês*
in a determinate way, separately: *aphôrismenôs*
in a unified way: *heniaiôs*
in a verbal way: *logikôs*
in every case, absolutely: *pantôs*
in general, universal: *katholou*
in general: *holôs*
in particular, peculiarly, by itself, specifically: *idiai*
in relation to each other: *prosallêlos*
in the proper (primary) sense, properly (speaking): *kuriôs*
inconceivable: *anepinoêtos*
indefiniteness: *aoristia*
indication, evidence: *tekmêrion*
indivisible, indivisible species, individual: *atomos*
indivisible: *adiairetos, ameristos*
inductive proof: *pistis epagôgikê*
inequality: *anisotês*
inexhaustible, never-failing: *anekleiptos*
infinitely powerful: *apeirodunamos*
inseparable: *akhôristos*
inspired: *epoptikos*
insubstantial: *anousios*
intellect: *nous*
intellective, intellectual: *noêros*
intelligence: *phronêsis*
intelligible: *noêtos*
intention: *prothesis*
intertwining: *diaplokê*

justice: *dikê*

kind, class, genus: *genos*
knowledge itself: *autognôsis*
knowledge, cognition: *gnôsis*
knowledge, theory, study: *theôria*

lack of clarity: *asapheia*
leading up to: *anagôgos*
lecture: *akroasis*
lifeless: *azôs*
limit: *peras*
limiting: *apoperatôsis*
line: *grammê*
link: *sundesmos*
love (Empedoclean principle): *philia*
love to know: *philotheamôn*

magnitude itself: *automegethos*
magnitude: *megethos*
make a statement: *apophainesthai*
malevolence: *panourgia*
many, the, multiplicity: *polla, ta*
material: *hulikos*
mathematical object, science: *mathêma*
mathematical objects: *mathêmatika*
matter known: *theôrêma*
meridian: *mesêmbrinos*
method: *methodos*
minor premise: *proslêpsis*
mobile: *kinêtos*
moment: *hôra*
monad: *monas*
monadic: *monadikos*
moved by others: *heterokinêtos*
movement itself: *autokinêsis*
multiplicity, the multiple: *plêthos*
musical: *mousikos*

natural number: *arithmos phusikos*
natural, natural philosopher: *phusikos*
nature, natural order: *phusis*
nectar: *nektar*
negation: *apophasis*
negatively: *apophatikôs*
non-existence: *anuparxia*
non-existent: *anupostatos*
non-generative, sterile, infertile: *agonos*
(in a) non-temporal way: *akhronôs*
nonsense: *teratologia*
notion: *ennoia*
number: *arithmos*

object of desire: *orektikos*
object of opinion: *doxastos*
object of science: *epistêtos*
objection: *enstasis*
of equal rank: *isotimos*

of life: *zôtikos*
of the same kind, of like form: *homoeidês*
on the basis of belief: *pisteutikôs*
one, more: *henikôteros*
one, the, unity: *to hen*
one: *heis*
opinion, based on, on the level of opinion: *endoxôs*
opinion, belief: *doxa*
opportune (moment), the: *kairos*
oppose: *antidiairein*
opposing, opposed: *antikeimenôs*
opposite, to be opposed: *antikeisthai*
opposite: *antithetos*
opposition: *antithesis*
order of angels: *angelikos, taxis*
order, realm: *diakosmos*
order: *diakosmein*
order: *taxis*
otherness: *heterotês*
outspoken: *parrhêsia*
overturns, destructive: *anatreptikos*

partial: *merikos*
particular: *hekaston, to kath'hekaston*
part-less, without parts: *amerês*
peculiarly qualified, the: *idiôs poion*
per accidens: *kata sumbebêkos*
perceptible: *aisthêtos*
perceptive: *aisthêtikos*
perfect: *teleios*
perfective (cause): *telesiourgos*
person of intelligence: *phronimos*
philosophers of nature: *phusiologoi*
place: *khôra*
plausibility: *paramuthia*
portion: *morion*
power, force, potentiality, in potency: *dunamis*
powerful: *poludunamos*
precise, accurate: *akribês*
precisely a being: *hoper on ti*
precisely a one: *hoper hen ti*
premise: *protasis*
present: *proteinein*
primary productive (cause): *prôtourgos*
principle, first principle: *arkhê*
prior and posterior, the: *proteron kai husteron, to*
privation, by: *sterêtikôs*
privation: *sterêsis*
problem: *problêma*
proceeding: *proodos*
produce: *dêmiourgein*
production (of the universe): *dêmiourgia*
productive, efficient: *poiêtikê*
proof: *kataskeuê*
proper: *idios*
propose previously, give: *prokataballein*
protection: *phroura*
proximate, immediate, close: *prosekhês*
purpose, intention: *prohairesis*

raise problems: *aporein*
raising problems: *aporêtikos*
ranked with: *suzugos*
reason: *logismos*
reasoning: *dianoia*
receptive: *hupodokhê*
reduction (to the impossible): *apagôgê*
reflection: *emphasis*
refutation: *elegkhos*
related: *sungenês*
relation: *skhesis*
representation: *phantasma*
rules: *basilikos*
ruling, originative: *arkhêgikos*

salvation, saving: *sôteria*
same, itself: *hauto*
science, knowledge: *epistêmê*
scientific: *epistatikos*
self-consistency: *sumphônia*
sense-perception: *aisthêsis*
separable, separate: *khôristos*
separative: *diakritikos*
signify: *sêmainein*
similar: *homoios*
similarity: *homoiotês*
simply, absolutely: *haplôs*
sophist: *sophistês*
soul: *psukhê*
sound, word, speech: *phônê*
speak falsehood: *pseudesthai*
specialist, expert: *tekhnitês*
specification: *prosdiorismos*
speech, statement, discussion, account, argument, ratio, reason-principle: *logos*
spontaneously: *automatôs*
state: *hexis*
state: *politeia*
statesman: *politikos*

Strife (Empedoclean principle): *neikos*
study, discussion, lecture: *skholê*
subaltern: *hupallêlos,*
substance, being, essence: *ousia*
substance, with regard to, in the very essence: *kat'ousian*
substantial: *ousiôdês*
substrate, what underlies, subject-matter, object: *hupokeimenon*
superabundance: *periousia*
supposition, opinion: *hupolêpsis*
sustaining: *sunokhikos*

teachable, teaching: *didaskalikos*
term, determination, definition: *horos*
testing: *peirastikos*
the 'that': *hoti, to*
the 'what': *ho ti, to*
the 'why': *dioti, to*
theologian: *theologos*
theoretical: *theôrêtikos*
thesis, position: *thesis*
thing, particular : *tode ti*
thing, reality, object: *pragma*
things in the universe: *enkosmia*
thinking, by, conceptually: *epinoian (kat', eis)*
thinking, intellection: *noêsis*
transcending: *exêirêmenos*
treatise: *pragmateia*
triple: *anatriploun*
true in both parts, to be: *sunalêtheuein*
true to Plato: *platônikôs*
trustworthiness: *pistis*
truth itself: *autoalêtheia*
two, dyad: *duas*

ultimate: *apoteleutêsis*
unconfused: *asugkhutos*
unco-ordinate: *akatataktos*
undemonstrated: *anapodeiktos,*
underlie (pass.): *prohupoballein*
understand, accept, include: *paralambanein*
unequal: *anisos*
ungenerated: *agenêtos*
unhesitatingly: *adistaktôs*
unification, unity: *henôsis*
uninvestigated: *adiereunêtos*
universal reason, (rational) principles: *katholou logoi*
universe, the: *pan, to*
univocally: *sunônumôs*
unlimited: *apeiros*
unmoved: *akinêtos*
unparticipated: *amethektos*
unprejudiced: *aprolêptos*
unstintingly: *aphthonôs*

validity (in refuting): *anaskeuastikos,*
vantage-point: *periôpê*

well (said), very: *hupereu*
well-being: *euphrosunê*
whole, universal: *holos*
wisdom: *sophia*
without cause: *anaitios*
without finality: *askopos*
word, name: *onoma*
words: *lexis*
world, ordering of the world, organisation, structure: *diakosmêsis*
world, the, universe: *kosmos*
worthy: *semnos*

zodiac: *zôidiakos*

Greek-English Index

This index lists a selection of the more important words, either from a philosophical or philological perspective, occurring in the Greek text. The translations given here may not always correspond to the rendering of them in a particular passage in the English text, since the demands of idiomatic translation may call for variation; but it should always be possible to work out what word is being translated.

Index of Aristotelian and Platonic Passages

References, in bold type, are to the page and line numbers of Kroll's edition, which appear in the margins of the Translation.

Subject Index

This index consists primarily of proper names mentioned in the introduction and text. Most major items of philosophical or philological interest are in fact covered in the Greek-English Index, but a few topics have been included here. References are to the pages of this book.

www.ingramcontent.com/pod-product-compliance
Ingram Content Group UK Ltd.
Pitfield, Milton Keynes, MK11 3LW, UK
UKHW020146290926
15471UKWH00004B/720

9 781472 558176